UNDENIABLE

How Biology Confirms Our Intuition That Life Is Designed

DOUGLAS AXE

HarperOne
An Imprint of HarperCollins*Publishers*

Pages 289–90 constitute a continuation of this copyright page.

HarperCollins books may be purchased for educational, business, or sales promotional use. For information, please email the Special Markets Department at SPsales@harpercollins.com.

FIRST EDITION

Designed by Terry McGrath
Illustrations credits on pages 289–90.

Library of Congress Cataloging-in-Publication Data has been applied for.

ISBN 978-0-06-234958-3

16 17 18 19 20 RRD 10 9 8 7 6 5 4 3

For Anita,
who showed me that one plus one
can be much greater than two

CONTENTS

CHAPTER 1

THE BIG QUESTION

In August of 2013, as I was making my way down a picturesque Cambridge street called King's Parade, I nearly collided with renowned British scientist Sir Alan Fersht. We were a short distance from Cambridge University's Gonville and Caius College, where he serves as Master among a distinguished group of scholars including the well-known cosmologist Professor Stephen Hawking. Fersht was exiting a shop, stepping across the sidewalk to his bicycle, and that was where our paths crossed.

I know him as Alan. I had been friends with him for so long—having worked at research centers directed by him from 1990 to 2002—that I assumed we probably still were friends, eleven years after we went our separate ways. Events had tested the friendship, though. From my perspective, an honest conversation making it clear where we stood with each other and why our working relationship had to end so abruptly would have been very helpful when I left his Centre for Protein Engineering in 2002. I had regretted the absence of that conversation over the years, and now, in the space of a few minutes, it occurred to me that he might have regretted it too.

Our time was short. I had a vacationing family waiting for me and Alan had a college waiting for him, so we settled for something less than closure. We did what we could do with a few minutes. After all that had happened previously, those few minutes reaffirmed our friendship, which was a good start.

The initial awkwardness of that encounter proved well worth enduring, as is often the way with awkwardness. I speak as something of an expert on the subject. Most people find their place in the stream of life early on by mastering the art of "going with the flow," but I seem to be one of the exceptions. I never set out to oppose the stream. Still, I found myself compelled to take a course you would never choose if the power of the stream were at the forefront of your mind. As anyone who's tried wading across swift waters knows, awkwardness was bound to follow.

I recall a question on a final exam near the beginning of my graduate studies at Caltech: *Which of the biological macromolecules is apt to have been the first "living" molecule, and why?* If that sounds like Greek to you, relax. I promise to write in plain English. All you need to know is that the question is about how life began, posed with the unstated assumption that it began by ordinary molecular processes. That assumption had been ingrained in biological thinking for so long that it went without saying. Every student in the class understood this, but I understood it more critically than most did. I knew the expected response to the test question, but through my critical lens, that response seemed scientifically questionable. So I had a choice: Do I go with the flow, or do I push against it?

I decided to give the expected answer in full and then—for extra credit—to state why I found that answer unconvincing. I explained why, contrary to the consensus view, I didn't think *any* molecule has what would be needed to start life. As shrewd as that seemed at the time, I learned when my exam was returned (with points deducted) that we students were expected not only to know current thinking in biology but also to accept it without resistance. We were there as much to be acculturated as educated.

I had learned my lesson. The stream of scientific consensus flows with an almost irresistible current.

Almost.

Awkward Science

Of all the controversial ideas to come from modern science, none has brought more awkwardness than Darwin's idea of evolution through natural selection. We know natural selection means "survival of the fittest," which in one sense isn't at all controversial. Indeed, Darwin's observation that fitter individuals are apt to have more offspring is so obvious it hardly needs to be stated. But how can something with so little content—a *truism*—possibly explain the astounding richness of life?

The biggest question on everyone's minds has never been the question of survival but rather the question of origin—*our* origin in particular. *How did we get here?* Even if you think natural selection is the answer, you have to admit to a degree of internal conflict over the matter. Francis Crick acknowledged this conflict, at least implicitly, when he cautioned that "biologists must constantly keep in mind that what they see was not designed, but rather evolved."[1] So if Darwin's claim is true, then it's a truth we all find ourselves doubting—at least subconsciously—and if it's false, then we're to be commended for doubting it. Awkwardness clings to it either way.

In fact, though you won't see this in any textbook, Darwin implicitly conceded something that adds to the unease surrounding his theory. All six editions of his book *On the Origin of Species* include a few paragraphs in the conclusion where he addressed the widespread rejection of his theory by his scientific peers. He began with a question: "Why, it may be asked, have all the most eminent living naturalists and geologists rejected this view of the mutability of species?" The answer, he thought, was their closed-mindedness. Sensing little hope of opening more than a few of those minds, he decided to "look with confidence to the future, to young and rising naturalists, who will be able to view both sides of the question with impartiality."[2]

To Darwin's own great surprise, this near total rejection of his theory turned to near total acceptance within just a few years. Up to the publication of the fifth edition of his book in 1869, his original gloomy assessment of the reception of his work wasn't in need of revision. Then in 1872, a mere three years later, the sixth edition followed those original paragraphs with this commentary:

> As a record of a former state of things, I have retained in the foregoing paragraphs, and elsewhere, several sentences which imply that naturalists believe in the separate creation of each species; and I have been much censured for having thus expressed myself. But undoubtedly this was the general belief when the first edition of the present work appeared. I

> formerly spoke to very many naturalists on the subject of evolution, and never once met with any sympathetic agreement. It is probable that some did then believe in evolution, but they were either silent, or expressed themselves so ambiguously that it was not easy to understand their meaning. Now things are wholly changed, and almost every naturalist admits the great principle of evolution.[3]

What would cause such a sudden reversal of scientific opinion? Did a new scientific discovery appear in the late 1860s or early 1870s—potent enough to convince the skeptics that Darwin was right after all? Clearly not, as Darwin surely would have cited such a decisive finding. But if science itself wasn't the cause of the change, then what was?

Whether he intended to or not, Darwin reveals here that *peer pressure* is a part of science, happening behind the scenes as the various scientific interests compete against one another for influence. If it's a plain historical fact that the experts didn't side with Darwin in the early 1860s, then why would he have been "much censured" by his peers for saying so? It's as though his colleagues wanted all mention of opposition expunged from the record now that this opposition had faded. Darwin resisted the pressure applied to him on that occasion, but what if others, perhaps under even greater pressure, were less able to resist? Might the earlier inability of some scientists to express their support of Darwin's theory—the silence and ambiguity of expression Darwin referred to—have been the result of peer pressure too? And if so, then might the sudden change in Darwin's favor have been more like a change of power than a change of minds—a sudden reversal of the stream's flow?

We have good reason to consider this possibility. The question of what controls the stream—why it flows this way and not that, and why it changes when it does—is every bit as important now as it was back then. If yesterday's scientists were influenced as much by human factors as by data, wouldn't this be equally true of today's scientists? And if it is true, what does this mean for the received wisdom of our day, which holds the evolutionary view to be the only one worth taking seriously?

As we think more about how science works, we'll see that those rare people who oppose the stream are the ones to watch.

Heroic Misfits

Thankfully, every generation has had a handful of rebels who are compelled to do just that. A countercurrent of awkwardness flows from these misfits in refreshing waves. Among the most beautiful examples of this I've come across is a man named Thomas Nagel, a professor of philosophy at New York University. He's a highly unusual atheist, the author of a superb wave-making book titled *Mind and Cosmos: Why the Materialist Neo-Darwinian Conception of Nature Is Almost Certainly False.*[4]

By way of background, the flag that has flown for many generations over the academy of higher education is that of a broad school of thought known as *materialism.*[5] The meaning here isn't the common one (an obsession with flashy cars or expensive clothes) but rather the view that *matter*—the stuff of physics—underlies everything real. Even if they don't use this term, atheists tend to subscribe to the materialist view of reality, believing God to be a product of the human imagination, which they believe to be a product of material evolution. Theists, on the other hand, believe the reverse—that the material universe was brought into existence by God, who is not material. Both views accept the reality of the physical world, but one sees this as the *only* reality whereas the other doesn't.

People on either side of this divide might think constructive dialogue is hopeless because everyone on the other side has fallen prey to wishful thinking. In practice, however, I find that atheists are more inclined toward this. Atheists have a pronounced leaning toward *scientism*, which is the belief that science is the only reliable source of truth. It's entirely understandable, then, that belief in God might look to them like wishful thinking—as though people of faith have let their hearts overpower their heads. Although people of serious faith (myself included) know this to be a misconception, our holistic understanding of human belief and behavior certainly does include the heart along with the head. We fully acknowledge that emotion can get in the way of clear thinking, but since we see this as a very general condition of humanity, we would never offer it as a particular weakness of atheism, the way so many atheists offer it as a particular weakness of theism.

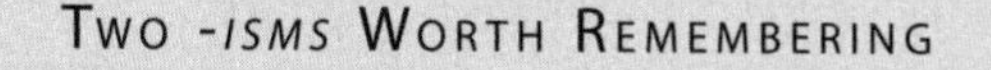

materialism:

the belief that physical stuff underlies everything real

scientism:

the belief that science is the only reliable source of truth

Returning to Thomas Nagel, as you may have guessed from the title of his book, he isn't your typical atheist. Most significantly, he roundly rejects the simplistic scientism that so many atheists still cling to. His atheism is heart-driven, and he isn't afraid to say so:

> I want atheism to be true and am made uneasy by the fact that some of the most intelligent and well-informed people I know are religious believers. It isn't just that I don't believe in God and, naturally, hope that I'm right in my belief. It's that I hope there is no God! I don't want there to be a God; I don't want the universe to be like that.
>
> My guess is that this cosmic authority problem is not a rare condition and that it is responsible for much of the scientism and reductionism of our time. One of the tendencies it supports is the ludicrous overuse of evolutionary biology to explain everything about life, including everything about the human mind.[6]

As a first-rate philosopher of the mind, Nagel actually changes the debate with this candid version of atheism. In light of his example, thoughtful atheists no longer have the luxury of assuming their worldview just works somehow—that dead molecules somehow formed simple life, and that simple life somehow formed us, despite all the apparent difficulties. Nor do they have the luxury of dismissing every argument against atheism on grounds of religious bias. Thoughtful theists, for their part, can no longer assume that atheism necessarily breeds contempt for faith.

Nagel is living proof that the awkwardness of bare-naked

honesty doesn't compare to the reward of engaging seriously the matters that concern us most—a principle that will serve us well as we begin our journey together. You need no special training to join this expedition. All you need is a healthy dose of curiosity and a healthy tolerance of the *good* kind of awkwardness—the kind that comes from challenging claims that ought to be challenged.

THE BIG QUESTION

Again, that one big question of our origin unites us—not because we agree on the answer but because we should all agree on the importance of *finding* the answer. Throughout history, it has been the foremost question of people searching for understanding: *What is the source from which everything else came?* Or, to bring it closer to home: *To what or to whom do we owe our existence?* This has to be the starting point for people who take life seriously—scientists and nonscientists alike. We cannot rest without the answer, because absolutely everything of importance is riding on it. To know where everything came from is to know where *we* came from, and where we came from has everything to do with who we *are,* and who we are has everything to do with how we ought to *live.*

> **THE BIG QUESTION**
>
> *To what or to whom do we owe our existence?*

If all goes well, our journey in this book will take us to the answer. We'll know we have arrived when we have an answer that not only rings true but also distinguishes itself as the *one* answer that rings true. There should be no credible alternative.

A map will be helpful as we begin. My aim over the next four chapters isn't to answer the big question but instead to show where we should be looking for the answer. Chapter 2 will introduce the intuition that creates internal conflict in all of us by tugging against Darwin's claims. This *design intuition,* as we will call it, is the very intuition Crick wanted us to suppress. Chapters 3 and 4 will be a short account of the unexpected lessons I learned while seeking a scientific solution to this internal conflict. These lessons weren't about the proteins I was studying but about the people I interacted with along the way—really, about people in general. With those lessons in hand, we'll see in chapter 5 that the answer we're seeking is to be found not in technical science but in something much more familiar—something I call *common science.* There will be plenty of glimpses of technical science along the way, but all of these will be presented with the nontechnical reader in mind. In the end, we'll see that mastery of technical subjects isn't at all needed in order for us to know the answer to the big question. Common science will be perfectly adequate.

The next section of the book—chapters 6 through 9—will be a journey through the important aspects of common science. The point of chapter 6 will be to provide a better understanding of what life is and what it isn't, which will prove helpful as we progress to the matter of where life came from. Chapter 7 will be a common-science refutation of the idea that natural selection explains how life came to exist in its countless remarkable forms. With natural selection off the table, chapter 8 will be an exploration of *searching*, showing that the many inventions needed for new life forms to evolve would have had to be *found* accidentally. Chapter 9 will finish the section by showing why invention *can't* actually happen that way. The intuition that Crick wanted us to suppress will end up being confirmed instead.

But all this only tells us what the answer to our question *isn't*. To arrive at a satisfying understanding of what the answer *is* will require us to continue our journey a bit further. In chapter 10 we will revisit the question of what life is, viewing it this time through the lens of invention. The following two chapters, 11 and 12, will serve as a reality check, first by considering carefully whether we have overlooked anything in rejecting the evolutionary explanation of life and then by asking whether the scientific community's defense of evolution looks more like a "science thing" or a "culture thing." Finally, chapters 13 and 14 complete our journey. There we examine the nature of life and humanity more deeply—leading to a clear picture of what the answer to the big question is—after which I offer a glimpse of what I hope biology will look like in the not-too-distant future, after a great many people join us on this journey.

CHAPTER 2

THE CONFLICT WITHIN

In 1986, not long after I challenged the assumptions behind that exam question at Caltech, I had a career-changing *aha* moment during a biochemistry lecture. Earlier, as an engineering student at Berkeley, I had learned about something called a *feedback loop.* The basic idea is simple, though considerable ingenuity is often needed to implement it effectively. Consider a familiar example: the thermostat used to control the temperature in your home. Factors like the weather outside or your cooking a meal inside often work against your aim of keeping the house at a comfortable temperature. The job of the thermostat is to counteract those disturbances by constantly measuring the indoor temperature and activating the heating or air-conditioning as needed. So the measured temperature is used as real-time information (feedback) by an automatic decision-maker (the thermostat) to control the very thing that is being measured: the temperature.

As straightforward as that sounds, it becomes much more complicated when highly active and complex processes must be kept under control. Knowing that the chemistry occurring inside growing cells is just that—highly active and complex—I was astonished when my biochemistry instructor revealed the elegance of the automatic decision-makers working *on the molecular scale* to keep the various chemicals of life at the right levels! The connection to engineering was so strikingly and delightfully obvious to me that I felt laughter well up.

As if anticipating my reaction, the instructor was quick to attribute these ingenious molecular decision-makers to unguided evolutionary processes. His message was clear: however remarkable these molecular control systems may be, they should be considered nothing more than natural accidents—just like everything else in biology.

Huh?

I didn't believe him. I knew—intuitively, anyway—that no string of accidents could possibly be so clever. At the same time, I sensed the weight of scientific authority standing with his interpretation and against mine. Notice I use the word *authority* here instead of *evidence*. He was the professor; I was the student. He could have filled the room with distinguished colleagues who agreed with his view, whereas I didn't even know any *students* who agreed with mine. And yet, for all the claims I had heard in lectures and read in textbooks about the inventive power of Darwin's evolutionary process, I hadn't seen a convincing scientific basis for these claims. As far as I knew, no one had shown how the amazing things of life *could* be accidental inventions instead of deliberate ones.

I was aware, of course, of the mountain of books and technical papers in which the facts of biology were interpreted through the lens of evolution, and I knew that many people perceived this huge body of literature as the very documented evidence I was seeking. I, however, saw this mountain only as confirmation (if any were needed) that the evolutionary lens was dominant in the life sciences. After all, countless ideas have gained large followings and generated piles of books, but no one is naïve enough to think all these ideas must therefore be true. No, I was after evidence of another kind—the kind with the power to persuade people who aren't initially in agreement. Nothing in that mountain of evolutionary literature seemed to be that. Nothing took the views of Darwin-doubters seriously. I knew this because I was one.

So, as a Darwin-doubter, I started planning to do the work myself. Though I had to be willing to be proved wrong, my strong hunch was that the results of this work would reverse the stream of scientific consensus. The stream's flow had been reversed before, so I figured it could be reversed again. I knew there were risks, but my motive for proceeding was too strong to be ignored. The troubling contradiction between what the voice of scientific consensus was telling me and what the voice of my own intuition was telling me—the *conflict within*—had to be resolved. That's exactly what I set out to do.

I have something even bigger in mind for this book, though. Here I hope to resolve the same conflict for *you*. It exists in all of us to some degree. We share it to the extent we share the intuition that life can't be an accident. And for all of us, *understanding* is what eliminates the contradiction. Technical understanding can be overwhelming for many of us, though, so while I will offer glimpses of what I consider to be the decisive technical science, I won't turn this into a science lecture. Instead, common science will be the thread that holds everything together.

To prepare us for that, let's start with an experiment performed in the kitchen rather than in the lab.

Starting with Soup

A team of researchers in the culinary sciences recently discovered a revolutionary new soup they call *oracle* soup, referring to the oracles (mysterious revelations) the ancient Greeks sought from their gods. Indeed, had this soup been known in the days of Homer, it surely would have been attributed to a powerful god. It looks just like alphabet soup—thin broth with little pasta letters and numbers swirling around—but this "soup of the gods" distinguishes itself by what it does, as this experimental recipe shows:

1. Fill a large pot with oracle soup.
2. Cover the pot, and bring the soup to a boil.
3. Remove the pot from the heat, and let the soup cool.
4. Lift the lid to reveal complete instructions for building something new and useful, worthy of a patent—all spelled out in pasta letters.
5. Repeat from step 2 as often as desired.

You don't believe a word of this, of course, and that's precisely my point. This was actually a *storytelling* experiment instead of a kitchen experiment. You were my experimental subject (sorry about that), but now I want you to examine the result. What did you observe? Well, in the space of a moment or two, you decided with complete confidence that oracle soup can't be real—you and everyone else who reads the account.

Interestingly, though, despite our collective certainty on this matter, most of us struggle to explain *how* we know oracle soup can't be real. Our explanations tend to be nothing more than restatements of our conviction that soup simply can't do such things. Children are content with those assurances, but adults surely ought to be able to do better. What makes us so sure oracle soup isn't real, then?

To ask a related question, how would we make sense of oracle soup if it *were* real? If we reflect on that for a moment, I think we would agree that no ordinary explanation would seem adequate for something so extraordinary. But if this is true, how can the evolutionary explanation of life not provoke that same skepticism? According to Darwin, each form of life owes its existence to a long succession of accidents—small mistakes of the kind that just happen from time to time. Anyone desiring a more lofty view of life can attribute these accidents to God if they wish, but Darwin's point, defended by evolutionary biologists to this day, is that no one *has* to do so. However skillfully the brush of natural selection appears to have picked hues from the palette of genetic mutations and applied them to the canvas of life, there's no need to think a personal hand ever guided that brush. We might just as well believe God guides each raindrop as it falls to the ground. The fact is, raindrops form and fall in accordance with certain well-known laws of physics, so nothing about rain ought to make anyone uncomfortable with leaving it at that. Rain happens. Life happens.

Rain comes from clouds, whereas life, according to Darwin's speculation, originally came from soup. Not from oracle soup but from *primordial* soup—the "warm little pond" Darwin described in a letter to his friend Joseph Hooker in 1871.[1] But if my claims about oracle soup were suspiciously extravagant, it's hard to see how Darwin's claims about primordial soup can avoid similar suspicion. To believe in primordial soup is, after all, to believe that a pool of mineral water set a process in motion that ultimately produced not just the genetic instructions carried by every form of earthly life but also innumerable marvels that go well *beyond* mere instructions—actual working wonders, like brains and compound eyes and adaptive immune systems and submicroscopic molecular machines, to name just a few.

In other words, the most peculiar aspect of Darwinism isn't that it takes credit for things that seem too extraordinary to be explained but rather that the explanation offered seems too ordinary for the job. The account of oracle soup is peculiar only in the first respect, and that was enough for us to dismiss it. Our skepticism would surely persist even if we witnessed a demonstration of oracle soup in action, because we would still find it easier to dismiss the demonstration as a clever trick than to accept the idea of a mysterious power working in soup. Only if oracle soup managed to stand up to all attempts by expert skeptics to debunk it would we reluctantly accept the idea that a mysterious power really *is* at work—carefully assembling messages with the pasta letters. Indeed, it's hard to imagine how else we would come to terms with the evidence. What's absolutely certain is that we would never accept the ordinary causes of physics and chance as explanations, because those causes are so clearly inadequate.

To be clear, I'm not suggesting that the falsity of the oracle soup story justifies rejecting the primordial soup story. There are obvious differences between the two, which we will examine in due course. For the moment, I'm simply saying that since we apply the same intuition to all accounts of remarkable occurrences, we shouldn't be surprised to find that the evolutionary story seems counterintuitive at times, even to those who accept it.

Berkeley psychology professor Alison Gopnik described the challenge this causes for teachers of evolution in a recent *Wall Street Journal* column. "By elementary-school age," she wrote, "children start to invoke an ultimate God-like designer to explain the complexity of the world around them—even children brought up as atheists."[2] In fact, Deborah Kelemen, a psychology professor at Boston University, found that even highly trained scientists are unable to fully rid themselves of the innate impression that there is purpose underlying the living world. According to her, "Even though advanced scientific training can reduce acceptance of scientifically inaccurate teleological explanations, it cannot erase a tenacious early-emerging human tendency to find purpose in nature."[3] Whether her materialistic presupposition will stand up to scrutiny remains to be seen, but her observation clearly affirms the universality and power of this design intuition.

The Universal Design Intuition

As a scientist, I recognize the need for caution here. Intuitions are such slippery things that we can hardly give an adequate firsthand account of them, much less a general account for all of humanity. Thankfully, we can proceed with something much more modest. In a moment I'll give one plausible account of how we might quickly decide that some outcomes can't be explained as accidents. Whether the method I describe is the one we actually use is less important than whether it justifies our conclusions. Specifically, we want to know whether the intuition that makes us doubt Darwin's theory is sound. If the answer to this is *yes,* as I think our journey will confirm, then Darwin's theory is in trouble whether or not we ever have a fully satisfactory account of how intuitions work.

With that qualification, I think the intuition by which we immediately perceive certain things to be the products of purposeful intent is close to the idea that some things are too good to be true. This expression doesn't mean that good things can't happen; it means certain good things can't *just* happen. They never come out of thin air. They only happen if someone *makes* them happen. We apply this insight to get-rich-quick schemes, for example, because they portray financial success as though it requires no skill or effort, whereas experience tells us otherwise. This hints at a universal rule for deciding what can and can't be attributed to accidental causes, which I'll state as follows:

THE UNIVERSAL DESIGN INTUITION

Tasks that we would need knowledge to accomplish can be accomplished only by someone who has that knowledge.

In other words, whenever we think we would be unable to achieve a particular useful result without first learning how, we judge that result to be unattainable by accident.

Again, whether there is one standard way we reach these judgments is not crucial to what follows. We don't even have to decide yet whether the rule is correct as stated. For now, the important point is that we all reach these judgments, often unanimously, and this rule fits these judgments reasonably well. I use the term *universal design intuition*—or simply *design intuition*—to refer to this common human faculty by which we intuit design.

As we proceed, it will become clear that I have something more ambitious in mind. I intend to show that the universal design intuition is reliable when properly used and, moreover, that it provides a solid refutation of Darwin's explanation for life. We'll have to think beyond our most familiar intuitions to reach that conclusion, but familiar points of reference will remain in sight throughout the journey. If the destination can be reached that way, as I believe it can, then having reached it, you will be fully capable of leading others along the same path.

The design intuition is utterly simple. Can you make an omelet? Can you button a shirt? Can you wrap a present? Can you put sheets on a bed? Tasks like these are so ordinary that we give them little thought, and yet we weren't born with the ability to do them. Most of the training we received occurred so early in life that we may struggle to recall it, but we have only to look at a young person still in the training years to be reminded that all of us had to be taught. Whether we taught ourselves these skills or were taught by others, the point is that *knowledge* had to be acquired in the form of practical know-how. Everyday experience consistently shows us that even simple tasks like these never accomplish themselves. If no one makes breakfast, then breakfast goes unmade. Likewise for cleaning up after breakfast, for making the bed, and so on.

Of course, this is anything but new. Plutarch, a first-century Greek historian, captured the universal design intuition nicely in an essay called "Fortune" (meaning *chance*):

> But can it be that those things which are most important and most essential for happiness do not call for intelligence, nor have any part in the processes of reason and forethought? Nobody wets clay with water and leaves it, assuming that by chance and accidentally there will be bricks, nor after providing himself with wool and leather does he sit down with a prayer to Chance that they turn into a cloak and shoes for him.[4]

According to the design intuition, neither bricks nor shoes get made unless someone makes them. As familiar as this intuition is, it turns out to have huge implications for biological origins, because the claimed exceptions are so concentrated there. And what dramatic exceptions they are! Bricks don't get made until someone makes them (or today, until someone makes the machine that makes them), but somehow much more complex things, like dragonflies and horses, *did* get made without anyone making them, we are told.

If you think this riddle has a solution that leaves evolutionary biology intact, I hope to convince you otherwise before our journey ends. To whet your appetite for what's to come, spend a moment contemplating a striking contrast of complexity. At the very low end of the scale are the many simple, everyday tasks that require very little thought, like the making of a bed, but that we know from experience are never accomplished without someone working to accomplish them. These things are far too simple to fascinate us but evidently too complicated to be done by accident. This realization seems to justify our sense that nothing impressive ever *does* happen by accident. Far beyond such simple things are the pinnacles of human technology, like robots and communications satellites and smartphones, which we also know can't appear by accident. Finally, at the highest reaches of the complexity scale are the true masterpieces—things like hummingbirds and dolphins—all of them alive, all of them eluding our best efforts to understand them. Some technophiles like to think that human ingenuity will one day produce their equal, and good things will surely come from rising to that challenge. To me, though, speaking as a fellow technophile, those masterpieces look positively untouchable.

I aim to give you a better sense of what I mean by this later in our journey. The next step toward resolving the conflict within, however, will be to gain a better understanding of what this thing we call "science" really is. For that, we'll focus less on scientific questions than on the scientific culture within which these questions are raised and answered.

CHAPTER 3

SCIENCE IN THE REAL WORLD

Determined to resolve the conflict between the design intuition and Darwin's theory, I spent much of my spare time from 1988 to 1990—the end of my Ph.D. years—reading as much as I could about evolution. I wanted to know who else was wading against the stream, and I was encouraged to find a few impressive skeptics. Several of these skeptics had given talks at a symposium organized in Philadelphia back in 1966 under the eyebrow-raising title "Mathematical Challenges to the Neo-Darwinian Interpretation of Evolution": Marcel Schützenberger from the University of Paris; Stanislaw Ulam at Los Alamos National Laboratory; MIT's Murray Eden—people who aren't easily ignored. Their talks, transcribed and published the following year,[1] presented thoughts at various stages of refinement. These short papers lacked the weight of finished research projects, but to my mind they amply demonstrated the need for such projects. The very fact that serious scientists were thinking and expressing these anti-Darwinian thoughts was intriguing.

Slightly troubling was the fact that in the twenty-some years between the symposium and my reading of it, nothing as healthy as that gathering seemed to have occurred again. It was as though a monumental train of thought had somehow not been allowed to continue its course. A few brave books in those years had challenged the evolutionary story on scientific grounds, most notably Michael Denton's *Evolution: A Theory in Crisis,* published in 1985.[2] But the apparent lack of any gathering of scientists at an established scientific institution to carry this critique of Darwin's theory forward suggested to me that the scientific establishment was not at all in favor of it. What had happened in Philadelphia in the late 1960s seemed, anyway, not to be possible anywhere in the late 1980s. For me, this weird opposition, if that's what it was, made the opposed work even more attractive. On top of the obvious intellectual importance was the danger-sport-like adrenaline rush that comes from being a scientific renegade.

I was in. If opposition prevented me from openly stating my aim as I embarked upon this dangerous career path, then I would keep my aim private.

Tiny Designs

What I found most intriguing in the small body of technical literature that challenged Darwin's theory was the improbability of characters becoming arranged into long functional sequences by accident. This is exactly what made us so suspicious of oracle soup. There the characters were alphabetic letters, and the functional sequences were written instructions. We knew intuitively that the accidental arrangement of pasta letters into instructions is so fantastically improbable that it can't happen.

The same intuition—the design intuition—applies to functional sequences built from *any* kind of character set, from the zeros and ones of computer code to the hieroglyphs on the Rosetta Stone. Fascinatingly, the molecular underpinnings of life provide two more prime examples: *gene* sequences and *protein* sequences.

We will get to genes in a minute. As for proteins, these are the molecules responsible for most of the cellular activities of life. By rough analogy, if we liken a cell to a car, then the individual protein molecules within the cell are like the individual mechanical parts of the car—proteins are that crucial to life. Each protein molecule is a long chain of connected "characters" called *amino acids*. These amino acids are small molecules with standard connectors on both ends and a protruding part in the middle. The twenty natural amino acids differ only in these protruding parts, which I'll refer to as *appendages* (see Figure 3.1).[3] If the sequence of amino acids along a newly made protein chain has the right properties, the whole chain folds up automatically (or nearly so[4]) inside the cell to form a compact three-dimensional structure. Like wire sculptures made from single wires, proteins can take on a great many different shapes, but unlike a wire, most protein molecules have a single preferred folded shape, the details of which are crucial to its function. Just as the parts of a machine must be shaped correctly to do their various jobs, so it is with proteins.

The preferred shape of each protein turns out to be specified by the sequence of amino acids along the length of its chain. But that raises an interesting question: How do cells "know" what these sequences should be? The answer lies in genes and

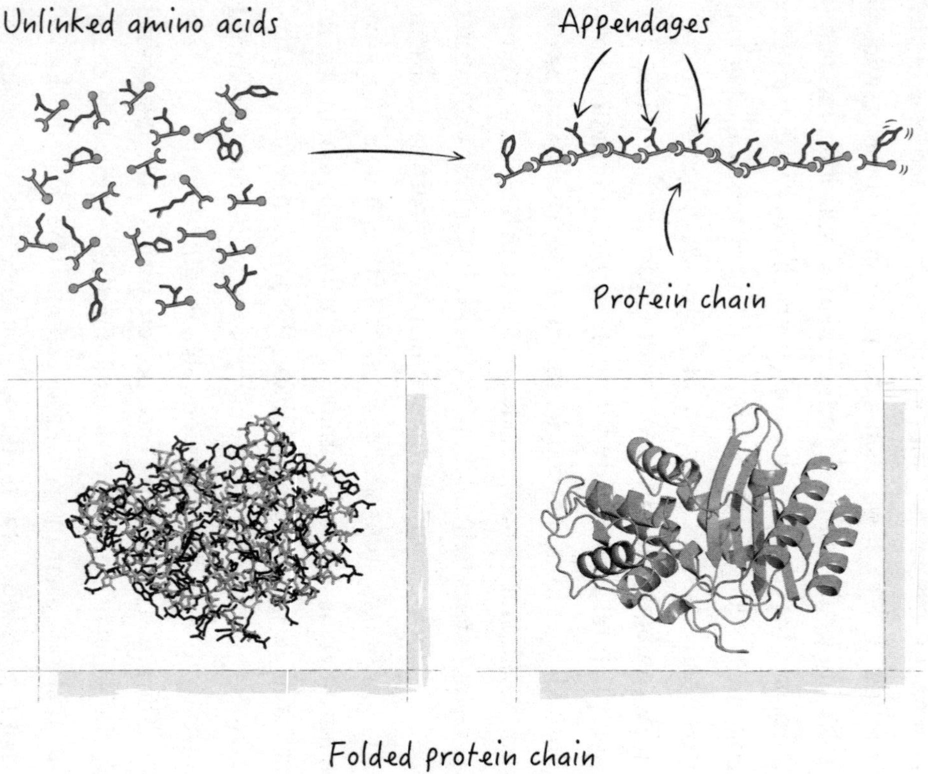

Figure 3.1 The construction of proteins from amino acids. Most living cells use the same basic set of twenty amino acids depicted in the upper left (with artistic license). Amino acids are linked one by one, in the precise sequence specified by a gene, to form a long, flexible chain-like molecule (*upper right*). The amino-acid sequences specified by most natural genes have the highly special property of causing the whole chain to fold into a well-defined three-dimensional structure, an example of which is shown in the lower left. Scientists use simplified representations to make it easier to see the features of these folded protein structures, the most common one being the "ribbon" diagram, shown for the same protein (called *beta-lactamase*) in the lower right. Each coil in a ribbon diagram represents an element of structure called an *alpha helix*, and each arrow represents a *beta strand*. These two elements make up most of the structures of all proteins, with the connections between the elements called *turns* or *loops*. Although the loops look floppy, like spaghetti, they usually have a firmly fixed structure just like the rest of the protein.

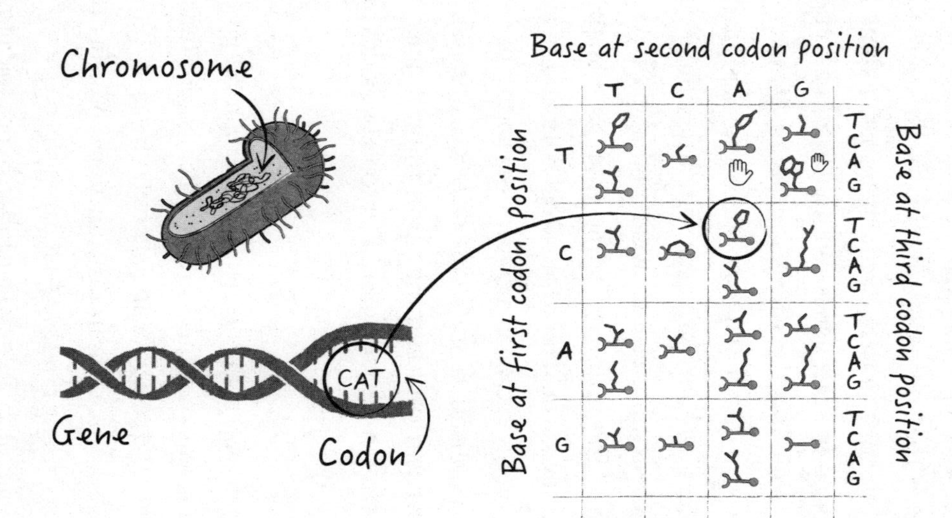

Figure 3.2 Genes and the genetic code that cells use to translate them. For our purposes, think of a gene as a stretch of chromosomal DNA. DNA chains are made from characters that, like amino acids, differ in their appendages. The DNA appendages are called *bases*. Because these bases come in only four kinds (represented by the letters *A*, *C*, *G*, and *T*), it takes a group of three consecutive bases, called a *codon*, to specify any one of the twenty amino acids.[5] A highly sophisticated molecular system involving about a hundred specialized proteins is used to interpret each of the 64 possible codon sequences as specifying one of the twenty amino acids (or the end of the protein chain, represented by the hand symbol). The end result is the set of codon "meanings" that we refer to as the *genetic code*, often represented in the form of a table, as shown.

the genetic code. Each protein molecule is constructed by linking amino acids according to sequence instructions carried by a gene. There's a trick to reading these genetic instructions, though. DNA consists of *four* types of characters joined in sequence, whereas proteins consist of *twenty* amino-acid characters joined in sequence. A code is therefore needed for cells to translate sequences of four into sequences of twenty. Life has precisely such a code: the famous *genetic code* that was cracked in the late 1960s (see Figure 3.2).

As we trace the source of proteins back, we see that the genetic code explains how the sequence instructions for proteins are encoded in their genes. But that raises another pressing question: *How did the various forms of life acquire these necessary genes in the first place?* Here our design intuition clashes with the scientific consensus, which attributes genes and proteins and everything else to accidental causes. As Michael Denton put it, "The intuitive feeling that pure chance could never have achieved the degree of complexity and ingenuity so ubiquitous in nature has been a continuing source of scepticism ever since the publication of [*On the Origin of Species*]."[6]

This skepticism kept coming up in the critiques of evolution I was reading in the late 1980s. Denton's book described the problem this way:

> There are, in fact, both theoretical and empirical grounds for believing that the *a priori* rules which govern function in an amino acid sequence are relatively stringent. If this is the case . . . it would mean that functional proteins could well be exceedingly rare. . . . As it can easily be shown that no more than 10^{40} [1 followed by 40 zeros] possible proteins

> could have ever existed on earth since its formation, this means that, if protein functions reside in sequences any less probable than one in 10^{40}, it becomes increasingly unlikely that any functional proteins could ever have been discovered by chance on earth.[7]

Simply put, it seemed likely to Denton that protein science was poised to disprove Darwin. I agreed, and I wanted more than anything to do this science.

Within a few years, my pursuit of that ambition took me to Cambridge, England. Working first in the Department of Chemistry at Cambridge University, I soon came to realize that opposition was not the only thing that had kept scientists from settling the matter Denton and others had raised. The kinds of experiments that were needed were easy to describe in theoretical terms, but they turned out not to be so easy to nail down in practical terms. The basic idea was to put Denton's claim that "functional proteins could well be exceedingly rare" to a decisive test. Doing this would require more experience and more careful thought.

In pursuit of this experience, I eventually landed at another major research center in Cambridge, this one having a rather extraordinary history behind it.

THE HUMANNESS OF GENIUS

Housed within an unimpressive box-like building at the south end of Cambridge was the highly impressive Laboratory of Molecular Biology—the *LMB*. Within months of opening its doors in 1962, the LMB could boast *three* Nobel Prizes shared among its scientists. Fred Sanger was the sole winner of the chemistry prize in 1958 for his discovery of the amino-acid sequence of insulin. The second and third prizes both came in 1962, one going to James Watson and Francis Crick for discovering, along with Maurice Wilkins, the double-helix structure of DNA, and the other going to Max Perutz and John Kendrew for discovering the first protein structures. Many more LMB Nobel laureates have been named since then, but the intellectual thrust that propelled the lab into high orbit traces back to the explosive success of that small initial group of people, originally headed by Max Perutz.

In September of 1999, I paid a visit to an office at the LMB that reflected the character of the man who occupied it, humble and tidy. Max stood opposite me, slightly hunched, using a wood lectern to support himself. Pain in his back made sitting difficult. Indeed, his body showed all of his eighty-five years, but his mind and his work schedule were those of a much younger man. Although he had long since handed leadership of the LMB

Figure 3.3 The "brick box" that served as the home of the Medical Research Council's Laboratory of Molecular Biology (MRC LMB) from 1962 to 2013.

Figure 3.4 Six Nobel Prize winners at the Nobel ceremonies in Stockholm in 1962, four of them associated with the research groups that formed the LMB that year. Shown from left to right are Maurice Wilkins, Max Perutz, Francis Crick, John Steinbeck, James Watson, and John Kendrew. Crick and Watson, both associated with the groups that formed the LMB, shared with Wilkins the prize for physiology or medicine, while Perutz and Kendrew—both LMB leaders—shared the prize for chemistry. Steinbeck received the prize for literature.

over to others, he continued to walk the halls almost daily, keeping abreast of the latest research and even contributing to a project here and there.

On his lectern were twenty-six sheets of paper that represented more than a year of my work. I was taking a calculated risk. A prevalent idea at the time was that proteins were not particularly fussy about the sequence of amino acids along their chains, and even less fussy about the identities of the amino acids that end up on the outside of their folded structures. According to many scientists then, all a protein needed in order to fold was an appropriate placement of water-loving and water-repelling amino-acid appendages along the chain. About five of the twenty appendages can be classified as water-repelling and seven or so can be classified as water-loving (the rest fall in between), so you can see how this simplified view would, if correct, make it much easier for evolution to find amino-acid sequences that fold to form new protein structures (Figure 3.5). In effect, the difficulty of arranging twenty kinds of appendages into a stable structure would be reduced to that of arranging just three kinds of appendages: water-loving, water-repelling, and ambivalent.

My paper opened by connecting this simplified view to work Max had done in the late 1960s. I knew this connection could backfire, though, because the rest of the paper described

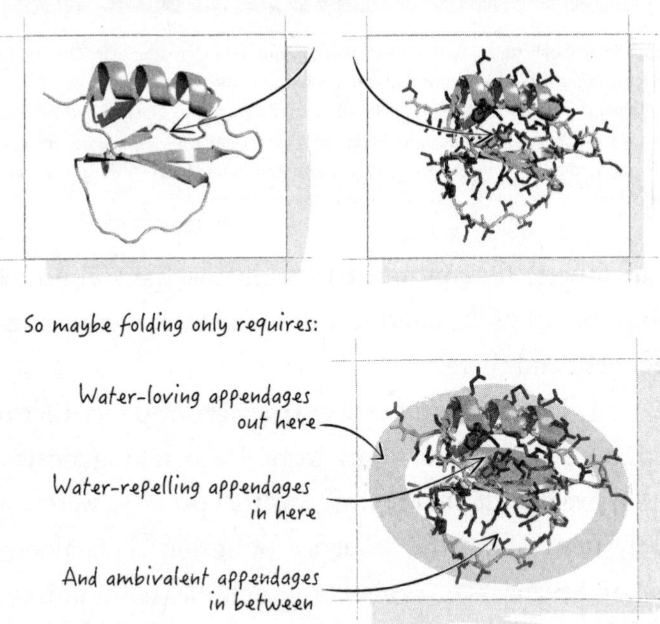

Figure 3.5 The simplified view of protein structure formation. All three images depict the same small protein, called chymotrypsin inhibitor 2. The sandwich-like packing of an alpha helix onto a group of beta strands (called a *beta sheet*) is shown with the appendages that stabilize this structure visible (*right*) or hidden (*upper left*). The tumbleweed appearance with the appendages is deceptive. Although you can see through the protein in this stick representation, a surface representation (peek ahead at Figure 7.5 for an example) would show that water can't enter the interior. The importance of excluding water from the interior by placing water-repelling appendages there was what led to the simplified view.

experiments that clearly showed the simplified view to be incorrect. Even I found this result surprising. My experiments had been performed on two different *enzymes*—the general term for proteins that perform specific chemical transformations. Having shown in 1996 that a particular small enzyme continued to do its chemistry even after all its interior amino acids had been randomly replaced with water-repelling alternatives,[8] I had assumed the exterior would be content with any combination of water-loving amino acids.

This turned out to be untrue. Shortly after I started the work, it became clear that both of the enzymes I was testing were completely inactivated after just a fraction of their exteriors were replaced in this haphazard way. I responded by redesigning the experiments, carefully replacing exterior amino acids in groups of five or ten, not haphazardly but with alternatives that were the most similar. Again, both enzymes were ruined in the process, long before their exteriors had been fully replaced.

The fact that the amino-acid replacements were now very conservative made this a significant result because it contradicted the prevailing view so clearly. These two proteins were much more fussy about the identities of amino acids on their exteriors than I and most other scientists had assumed, and moreover, the method by which I had shown this suggested the same was true of other proteins. In short, I had shown that the ability of proteins to keep working after a small number of their amino acids are replaced—one of the main justifications of the simplified view—didn't mean that these changes were harmless. It only meant the harm had not yet reached the breaking point. The breaking point *is* always reached as more changes are introduced, even changes of the conservative kind I was using.

My hope during the week or so between handing my paper to Max and sitting down to hear his thoughts was that any unease this new finding might stir in him would be offset by his appreciation of the clarity of the result. If all went according to my plan, I would leave the room with a strong endorsement from one of the greatest scientists alive, which would surely pave the way for my paper to be published in an elite scientific journal.

This was not to be. I listened politely as Max, in a state of mild agitation, complained about things that, to my mind, had nothing to do with the substance of my work. The man I had hoped to impress was annoyed instead. "I was very disappointed with the level of my exchange with Max Perutz," I wrote to a friend a short time later. Thankfully, other experts viewed my paper more favorably. One of these, himself a prominent protein scientist at the LMB, described my findings as "both startling and convincing." So after passing the test of peer review, my paper was published in the *Journal of Molecular Biology* (*JMB*) in August of 2000.[9]

Much later, with the benefit of years of reflection, I came to a new understanding of my meeting with Max. As difficult as our interaction in his office had been for me, I began to realize that he had shown me something more important that day than anything I had been hoping to show him. What I learned will sound too obvious to be profound, and indeed, although it *is* obvious, this happens to be one of those obvious truths we easily lose sight of: Max Perutz, the small giant who deservedly occupies a position in the history of science well above that of most Nobel Laureates, was as *human* as you and I are.

Somehow, with the conferring of rare honors, with the establishment of scholarship funds and the dedication of buildings that bear a person's name, with oil portraits and marble busts and postage stamps bearing a person's likeness, with the passing of the person in the flesh and the growth of a legend to take his or her place—somehow the fallible aspects of humanness we most easily relate to evaporate, leaving us with an image that hovers midway between heaven and earth, neither divine enough to be worshipped nor human enough to be hugged.

Perhaps this tendency to idolize the legends of science is connected to a skewed view of the whole scientific enterprise. Many of us, including me, have bought into the idea that science, though practiced by humans, has managed to rid itself of the human flaws that leave their mark on every other human undertaking. The purity of science is guaranteed by the rigor of "the scientific method," we think.

Astrophysicist Neil deGrasse Tyson described this utopian view as follows in the first episode of the *Cosmos: A Spacetime Odyssey* television series:

> This adventure is made possible by generations of searchers strictly adhering to a simple set of rules: test ideas by experiment and observation; build on those ideas that pass the test; reject the ones that fail; follow the evidence wherever it leads; and question everything. Accept these terms, and the cosmos is yours.[10]

That all sounds very nice. And if ideas could be tested with a meter, the way batteries and fuses can, then Tyson's simple rules would work. But if we intend to question everything, perhaps we should begin by questioning whether the human testing of human ideas can really be so simple, considering how complicated humans are.

Nowhere are these complications more evident than in the discussion of big ideas that touch the way we live, because here we find that *everyone*—scientists included—has a strongly held view. And the very biggest ideas are those that offer answers to the all-important question of how we got here. We should by all means trust the scientific community to tell us how many moons orbit Neptune or how many protons are packed into the nucleus of a cobalt atom. Why would anyone distort facts of that kind? Matters where everyone wants to see things a certain way, however, are a completely different story. With *those* we should always apply a healthy dose of skepticism.

Many of us have bought into the idea that science, though practiced by humans, has managed to rid itself of human flaws. But if we intend to question everything, perhaps we should begin by questioning whether the human testing of human ideas can be so simple, considering how complicated humans are.

From Utopian Science to Authoritarian Science

Having not yet come to appreciate this human factor fully, I was quite confident by 2000 not just that the scientific facts were at odds with the evolutionary story but also that with the right protein experiment I could reverse the stream of scientific consensus by *proving* so. I hadn't done this key experiment yet, but I knew how it would be done. Holding a utopian view of science very much like the one Tyson describes, I was convinced that no matter how startling a scientific result may be, no matter how many scientists may react with incredulity or how many textbooks may have to be rewritten, science always sides with the truth in the end. And maybe it does. But had I seen myself as less exceptional, recognizing that many other scientists had been offering similarly weighty challenges to Darwinism for well over a century, I might have reached the sobering realization of how *long* the scientific community can take to settle on the truth.

Oddly enough, I now see how the pursuit of prestige—so evident in my own life—goes a long way toward explaining how science gets stuck on certain wrong ideas. In the professional world of science, prestige is bestowed in the form of praise, and not just any praise but the rare praise of those who are themselves most highly praised. Knowing how fickle praise can be, though, why would anyone assume that praiseworthy science always gets the praise it deserves? The sight of true words eliciting a strongly negative response is familiar to everyone, in all walks of life. Why, then, would anyone believe that the road to scientific truth and the road to scientific prestige are one and the same?

The answer, I think, is that when we fall for the utopian view of science, truth and prestige do appear to be on the same road. If we assume scientists are single-mindedly driven by the quest for truth and nothing else, then we expect those scientists with the keenest perception of the truth to rise to the top.

These top-notch scientists form an elite body of experts whose consensus opinion is the surest indicator of the truth there is. Prestige and truth then seem inseparable, as though they are just two different names for the same destination. And we need only follow the road a short way toward this destination to see that it is also the road to *authoritarian* science. With the truth perceived to be so reliably in the hands of the elites, we ordinary folks need not concern ourselves with the details when the elites are challenged. Instead, we wait patiently for them to deliver their official response, which is sure to be correct, we assume.

Of course, as a challenger of the consensus view of biological origins, *I* had to concern myself with the details, but I was strangely confident that the challenge I was mounting would compel the scientific authorities to concede out of sheer inability to oppose the truth. Confidence can be a good thing, but in retrospect, I see that mine was tainted with pride, which is not a good thing. I say this because my intent in showing you the less flattering side of science is not to make me look good or others look bad, and certainly not to make science look bad. My purpose is instead to promote a realistic view of humanity and of science as a human undertaking. After all, we won't really love science until we learn to love *real* science—not a hypothetical pursuit in a utopian world but an intrinsically human pursuit in *this* world, however imperfect.

What happened next turned out to be just the right medicine for my pride, though not the sort I would have prescribed for myself.

CHAPTER 4

OUTSIDE THE BOX

Under the directorship of Alan Fersht, whom I mentioned in chapter 1, the Centre for Protein Engineering—or CPE, as we knew it—occupied a building that was joined by a connecting corridor to the brick box that housed the LMB. For the most part, the engineering done by the forty or so scientists at the CPE consisted of designing small alterations to natural proteins in order to study how their chains fold into compact structures. One project, though, had a much more ambitious form of engineering in mind.

While I was doing the work that led to the 2000 *JMB* paper,[1] a colleague of mine, Myriam Altamirano, was attempting to re-engineer a natural enzyme in order to make it perform the function of a different enzyme. Like many other scientists at the time, she was using a hybrid approach that combined aspects of design with aspects of evolution. In all of these projects, the idea was first to make informed guesses as to what parts of the original enzyme should be changed and how, and then, after implementing these changes, to use the standard laboratory version of evolution (mutate → select → repeat) to sort out any minor problems. Although this strategy could work in theory, the limitations have become increasingly apparent in the years since. Eleven years later, some of the leaders in the field conceded that "efforts to date to generate novel catalysts have primarily demonstrated that we are getting good at making bad enzymes. Making good enzymes will require a whole new level of insight, or new methodologies altogether."[2]

The crux of the problem is that the evolutionary step at the end accomplishes so little that success rests almost entirely on the ability to make the right guesses in the first place. But, of course, if we knew how to do *that,* the evolutionary step would be largely superfluous. In other words, evolution seems to be an inadequate replacement for knowledge. Indeed, if our design intuition proves true, *nothing* is an adequate replacement for knowledge.

Very good informed guesses, however, are tantamount to knowledge, and in this case Myriam's guesses seemed to be that good. She found that her evolved engineered enzyme worked as well as the natural enzyme it was designed to imitate—a remarkable feat in a field where the term "success" usually had to be applied very generously. After writing up her results, Myriam submitted her paper for publication in the prestigious journal *Nature* around the time I met with Max Perutz. Her paper passed *Nature*'s peer review and appeared in February of 2000.[3]

A Group Without a Leader

Hoping Myriam's strong success would pave the way for more success, as is often the way in science, several Ph.D. students began to work under her supervision on projects that extended her method to other enzymes. But technical challenges began to present themselves, and just as this was becoming evident, the students suddenly found themselves without a leader. In late 2001, Myriam unexpectedly left the CPE. As the director of the CPE, Alan knew he would have to find someone to take her place, and since his own work had never focused on protein evolution, he knew he would have to look to someone else. The rapid three-year timetable for completing a Ph.D. in the British system made the situation all the more urgent for the students who had been in her charge.

Knowing that my work was increasingly touching on protein evolution, Alan approached me. After stressing the significance of Myriam's work, underscored by her paper in *Nature,* he spoke with me about the students who had been stranded without Myriam, ending with words to the effect of "*You* know a lot about protein evolution, Doug." I knew he was offering to hand the leadership of this group of students to me, but I didn't see how I could accept. Although Alan saw that I was doing careful, critical work that touched on protein evolution, he didn't know the details of my current project, and he probably underestimated the degree of my evolutionary skepticism. Myriam's group was abuzz with the idea that evolution could work wonders, whereas I had come to the opposite conclusion. How could I lead a group of people who seemed to be heading in a direction opposite of mine?

The indirectness of Alan's offer enabled me to decline indirectly, which is not my usual style. In this case, however, I took the easy option. By not saying I would lead Myriam's group, I conveyed to Alan that I wasn't interested in leading her group without having to explain why.

Science with Ambition

I had been proceeding with caution for a long time. Even before that *aha* moment I recounted at the beginning of chapter 2, I had been a quiet critic of materialism. My views were known to close friends, but they went no further than that. Written declarations of my thoughts were restricted to the bulletin board in my student room, which was cluttered with personal proverbs. One was a higher version of the design intuition: "It is intuitively obvious to me that a mere collection of atoms cannot attain consciousness. It can never become aware of its own existence." Another supported that higher intuition with a simple argument. I deduced from the reality of human free will that humans can't be material things, and "therefore, man did not evolve from the physical."

I never intended to keep silent forever, though. My plan all along was to continue thinking and working carefully in the hope of solidifying my early design intuitions and eventually earning the opportunity to communicate them publicly. If and when that time came, I was sure that science would be the best podium from which to speak. My utopian version of science wasn't contradicted by anything I knew at the time. I understood that people had their biases, and I had seen the prejudicial attitudes of anti-faith scientists. But the scientific arguments in defense of the design intuition seemed incomplete to me, and because I felt certain a complete argument could be made, I held to the idea that this argument would be widely accepted.

Now, if you're wondering whether it's legitimate for scientists to hope for a particular result when they set their goals, I can assure you that it is. We do this all the time. The search for extraterrestrial intelligence is a well-known example. SETI involves the work of many scientists who hope their search will one day prove successful. They have no proof, but science never starts with proof. Like every other worthwhile undertaking, science starts with ambition. The same could be said of the many scientists who devote themselves to finding cures for various diseases. There is no proof that these long-sought cures will be found, but the goal and the ambition are there, and this is no small thing. Scientific proof never comes without those key ingredients.

> *Harm comes to science not by people hoping to find a particular result but by people trying to suppress results that go against their hopes.*

When we consider who has the power to suppress unwelcome results, we see right away that the view most likely to cause suppression is the majority view of the scientific community.

A PERFECT STORM

In early 2002, scarcely a month after I passed up Alan's offer to assume leadership of Myriam's group, there was discussion in the LMB cafeteria of a possible problem with the results reported in Myriam's *Nature* paper two years earlier. It sounded serious. A graduate student who had been sorting through the storage tubes in Myriam's freezer had found that the labels on certain critical tubes didn't match the contents, and most troubling of all, the re-engineered enzyme that had received so much attention seemed not to work. A sinking feeling descended on the whole lab at the thought of several graduate students having spent a year or more of their precious time on projects that were predicated on a mistake.

Within days of the first mention of the inconsistencies, the nightmare was confirmed. The storage tube investigation revealed that Myriam had found her enzyme to perform as well as the natural enzyme because it *was* the natural enzyme. As in all laboratory selection experiments, she had looked for signs of bacterial growth under conditions where growth can't occur unless the desired function is present. In this case, however, nothing *should* have grown on her petri dishes because neither her designed changes nor the subsequent mutational variations on those changes actually caused the desired function. By accidental cross-contamination, a few cells of the strain with the natural enzyme had been mixed in with the cells that couldn't grow. This meant her positive result was really a false positive. A brief note of retraction soon appeared in *Nature*, leaving no doubt as to the status of the paper that had been published in 2000: "We conclude that the results are unsound."[4]

As if these internal events were not creating enough tension, the intelligent design (ID) view of biological origins was beginning to make headlines in the UK in connection with escalating controversy over the teaching of alternatives to Darwinism in state-funded schools. As my boss, Alan knew I had been receiving fellowship and research money from the major funder of ID work, the Discovery Institute, for several years. He had never asked me to explain why Discovery was interested in my work, and being well aware that the connection to intelligent design could generate controversy, I never brought it up. Alan had earlier mentioned the Discovery Institute website, so I presumed he had made the connection himself and was not bothered enough by it to discuss it with me.

But the hostile treatment of ID by the British news media seemed to be having an effect on him. I was the first person in the lab one morning in February of 2002. Alan usually made his rounds through the labs later in the day when work was in full swing, but on this morning he dropped in early to have a word with me. He seemed tense. He approached me as if there were a pressing matter he needed to discuss, yet he seemed unable to initiate the discussion. I assumed what burdened him was the question of whether he could allow me to continue my work at the CPE, knowing that I was a part of this thing being portrayed so publicly in such awful, conspiratorial terms.

There's no easy way to initiate a conversation like that, but any other conversation would miss the point. And if my future at the CPE was indeed the point, the point was missed. After mentioning that he had just listened to a BBC radio program discussing intelligent design, Alan put a few questions to me, somewhat awkwardly.

"You know this William Dembski fellow, don't you?"

"Yes."

"And you know about his intelligent design theory."

"Yes."

"Tell me, then, *who is the designer*?"

That was the top question asked by critics of intelligent design back then. They thought the answer would expose deception on the part of ID proponents. Their underlying assumption was that ID proponents were being coy about the identity of the designer of life in order to construct a version of creationism that, by avoiding the G-word, could be taught in American public schools. In reality, the question only exposed confusion as to what ID is.

How to Spot a Fake ID

The truth is that ID and creationism have always differed fundamentally in their methods and starting assumptions. Creationism starts with a commitment to a particular understanding of the biblical text of Genesis and aims to reconcile scientific data with that understanding. ID, on the other hand, starts with a commitment to the essential principles of science and shows how those principles ultimately compel us to attribute life to a purposeful inventor—an intelligent designer. ID authors settle for this vague description not because they want to smuggle God into science but because the jump from "intelligent designer" to "God" requires something beyond the essential principles of science.

The confusion over ID really stems from broader confusion as to what these essential principles are. Intelligent design takes a minimalist view. If science is the application of reason and observation to discover objective truths about the physical world, then doing science requires accepting just a few things—none of them controversial. First, we must accept that objective truths exist, as we all naturally do. Then we must accept that some of these truths pertain to the physical world, and that some of those that do can be discovered through human observation and reasoning. Since we all engage in this discovery process from an early age, we all naturally accept these propositions.

There is nothing more.

In fact, adding anything to this essential set of propositions causes two serious problems. First, the resulting embellished definition of science excludes what shouldn't be excluded, namely any work that adheres to the essential set without adhering to the embellishment. For example, if a group of people were to insist that science can't be done properly without accepting that life exists on other planets, then that group will refuse to consider any work done from a contrary perspective, even though this work may be perfectly legitimate science. Worse, embellishments run the risk of pressuring scientists into accepting *wrong* answers by ruling the right answers "unscientific."

As odd as this situation may seem, it's not hypothetical. The scientistic view introduced in the first chapter—*scientism*—is the most striking example of an embellished version of science that has risen to prominence. The reason adherents to this version hold science to be the only legitimate source of truth is that they also hold to materialism. This commits them to the idea that there isn't anything but physical stuff, and because science is the only way to know the truth about physical stuff, this leads them to conclude that science is the only source of truth. The materialist commitment *itself,* though, is completely unnecessary to science and therefore a harmful embellishment.

Later in our journey we'll see how scientism unravels with the unraveling of materialism. Our design intuition will turn out to be good science, whereas scientism will turn out to be bad philosophy. For now, just notice that scientism makes itself vulnerable by hitching itself to materialism, which has no place in science.

The Ides of March

Alan's questioning didn't seem to lead anywhere on that February morning, but the mounting tension surrounding intelligent design in 2002, and the way this tension amplified the problems caused by the collapse of Myriam's result, left me thinking my time at the CPE might be coming to an end. If I could somehow become the solution to these problems, though, my position would become secure, I reasoned. I had passed up the opportunity to lead Myriam's students because our projects were pointed in opposite directions. Now that their projects were on the verge of being abandoned, however, the idea of salvaging them by viewing them in *reverse* seemed promising. In other words, if I could get Alan and the students to consider interpreting their results not as proof that converting enzymes to new functions was easy but as proof that it was *hard,* then I would be happy to provide the needed leadership. I proposed a meeting to discuss this idea. Alan was very receptive, as were the students, so the meeting was scheduled for the final week of February 2002.

My gross underestimation of the difficulty of getting people to change the way they think in an hour or two is humorous to me now, but none of this seemed funny at the time. With visuals prepared and the outline of my argument well rehearsed, I took up the challenge of convincing the Herchel Smith Professor of Organic Chemistry at Cambridge University, who also happened to be a Fellow of the Royal Society and the director of the Cambridge Centre for Protein Engineering, along with half a dozen graduate students that their view of protein origins was incorrect, and that the failed projects could be combined with my project to make a strong case for the correct view.

Needless to say, my pitch was not a smashing success.

Years later, an article in *New Scientist* magazine about Biologic Institute (titled "The God Lab"[5]) revealed that one of my fellow scientists at the CPE had been pressing Alan to dismiss me because of my connection to ID. The article says Alan refused to do so, quoting him as saying, "I have always been fairly easy-going about people working in the lab. I said I was not going to throw him out. What he was doing was asking legitimate questions about how a protein folded." According to the article, I left the CPE after "Axe and Fersht were in dispute with each other over the implications of work going on in Fersht's lab."

The truth is that Alan did, in the end, give in to the internal whistle-blower who wanted me removed, though I certainly accept his account of having resisted this for some time. When he did finally act, I interpreted the awkwardness of his action as an indication of his reluctance. There was no heart-to-heart conversation or even a word spoken face-to-face. When everyone gathered in the customary way to bid me farewell, Alan was conspicuously absent. All I received was an e-mail from Alan's assistant on the eleventh of March 2002, succinctly stating that the CPE was "very short of [lab] bench space" and declaring Alan's solution: "Please vacate as soon as possible and by the end of March latest."

Evidently the needed space matched my dimensions exactly, so, after saying my good-byes to everyone present, I said good-bye to the CPE and to the brick box at the other end of the corridor: the LMB.

Conscience and Courage

The truth is that I may well have made the same decision if I had been in Alan's position. After all, challenging the evolutionary story was *my* calling, not Alan's. *I* am the one who accepted the risk of pursuing research I knew would lead to a confrontation with the scientific establishment. Since I had never consulted Alan on that aspect of my direction, it would have been presumptuous of me to think he would be willing to shoulder some of that risk himself.

Further confirmation of the risks came from all directions over the course of that week in March 2002. Heading into the weekend, *The Guardian*, a major British newspaper, ran a story on Friday afternoon, March 8, that began with its particularly alarmist version of whistle-blowing:

> Fundamentalist Christians who do not believe in evolution have taken control of a state-funded secondary school in England. In a development which will astonish many British parents, creationist teachers at the city technology college in Gateshead are undermining the scientific teaching of biology in favour of persuading pupils of the literal truth of the Bible.[6]

That set a whole gaggle of whistle-blowers off, and within days Prime Minister Tony Blair was fielding questions on the matter in the House of Commons.[7] Meanwhile, on Monday and Tuesday, March 11 and 12, the New York newspaper *Newsday* ran a two-part series under the heading *Creation vs. Evolution*,[8] which included the following provocative reference to my work:

> In the meantime, intelligent-design advocates have pointed to a third line of research as "the most promising development in the next few years" and yet another potential roadblock to evolution. The research, by Douglas Axe of the Centre for Protein Engineering in Cambridge, England, introduces a concept called "extreme functional sensitivity" that relates a protein's specialized function to the changes permitted in its amino acid sequence. Axe's premises [sic] are hinted at in an article published two years ago in the *Journal of Molecular Biology,* but [William] Dembski and others say Axe plans to go public with his full findings soon and "shake things up."[9]

I had said nothing so provocative. *Newsday* reporter Bryn Nelson had asked me at the beginning of March whether I thought the results of my 2000 *JMB* paper meant that the enzymes I studied didn't originate by evolution. I was careful to restrict my response to what was presented in the paper:

> I don't think the data presented in the *JMB* paper allow one to draw this conclusion. That paper does reveal that the constraints imposed by function on sequence are unexpectedly high, and this raises some important questions that need to be explored further and which I hope to explore further.[10]

Indeed, as I had just disclosed to Alan and the graduate students, I *was* exploring those questions further, and I had a sense of where the results were pointing, but the last thing I wanted to do was compromise my research by talking about unfinished work with a reporter.

The media storm soon dissipated, and in the end, I was able to complete the project I described to Alan and the graduate students at the Babraham Institute, located just outside Cambridge. Like the prior study, this one was accepted for publication in *JMB,* appearing in August of 2004.[11]

The tension continued after the storm, though, and still continues today, as does the scientistic interpretation of this tension. According to this now-familiar view, people of faith who challenge Darwinism are *really* pushing religion, even if their challenge has a scientific look to it. That being so, we need to warn everyone not to be deceived by appearances. Blow your whistles! The religious agenda is the enemy that threatens science, so all enlightened people should defend science against this enemy, we're told.[12]

The real problem for science, however, is not people having agendas (as they always do) but rather the *institutionalization* of agendas. This is the embellishment problem we discussed earlier. Once an embellished view of science becomes established, active suppression of dissent becomes inevitable, with predictable consequences. Everything that opposes the institutionalized agenda is labeled "anti-science" by those working to protect the agenda, and the fear of that label quickly enforces compliance among the timid.

Something even greater than science is at stake here. To see this, we need to go back to the question we considered at the outset: *To what or to whom do we owe our existence?* Pondering this, we see that the most significant cost of giving in to the whistle-blowers is not violation of our sense of fair play but participation in the systematic devaluation of human life. It is the cost of remaining quiet as young people who innately know themselves to be the handiwork of a "God-like designer" are indoctrinated with the message that they are instead cosmic accidents—the transient by-products of natural selection.

University of Washington psychology professor David Barash brings the sanctioned message to two hundred undergraduates every year in his course on animal behavior. With professorial authority, he declares to his young captive audience, "The more we know of evolution, the more unavoidable is the conclusion that living things, including human beings, are produced by a natural, totally amoral process, with no indication of a benevolent, controlling creator."[13] His agenda, clearly, is to treat human behavior as just another example of animal behavior, all of which he thinks is ultimately explained by evolution.

If this explanation turns out to be untrue, then his indoctrination is a predictable tragedy. Barash believes the falsehood that was instilled in him when he was a student, and having believed it, he dutifully assumes responsibility for instilling the same falsehood in other young people. But the fact that his actions are predictable makes them no less destructive. Contemplate for just a moment the dystopian vision of a generation of human beings believing in their hearts that they are nothing more than bestial accidents fending for themselves in a world where morality is a fiction, and you begin to grasp the true stakes.

Heroes are badly needed here, and we have every reason to think they will take their conspicuous stand in each generation. After all, having whistles blown on you is a small price to pay for the privilege of defending the existence of moral truths. If you think these heroes need to have Ph.D.s, I hope to convince you otherwise in the next chapter. When it comes to defending the big question of our origin, everyone is scientifically qualified.

```
Date: Tue, 26 Mar 2002 12:11:31 -0500 (EST)
Subject: Re: query

Doug:
      Forgive me if I'm wrong, but I get the feeling -- call it 'body
language' -- that you are avoiding directly answering my questions. I
think you know what I'm asking. You have indicated in the past that your
work has nothing to do with ID, yet your affiliations and conversations
with Dembski seem to indicate otherwise. You can't have it both ways. Are
you sort of quietly, or unobtrusively, trying to get ID friendly research
into the literature, or are you not? Are you afraid that by being openly
supportive of ID your reputation will be besmirched among your colleagues?
Frankly, the fact that Dembski keeps mentioning your work even though you
claim no link to ID smells sort of fishy. So, what's going on?
BP

                         -------------
          We will first understand how simple the universe is
          when we recognize how strange it is.
                         John A. Wheeler
                         -------------
                        We miss you, Carl...


Barry A. Palevitz, Professor


Date: Thu, 18 Apr 2002 17:15:35 +0100 (BST)
Subject: Re: query

Barry-

I've been away for a couple of weeks.

In answer to your questions, I've been neither evasive nor inconsistent.

I'm open to the possibility of an evidence-based design argument in
biology, and that explains the connection to the Discovery Institute.  At
the same time, I haven't yet seen the evidence to justify such an
argument, and that explains why I haven't put such an argument forward or
defended arguments that others have put forward.

Like you, I'm well aware that preconceptions can color one's thinking.
Perhaps unlike you, though, I'm also aware that since we all have them,
we're all susceptible to their influence.  Sympathy toward design
arguments is no more capable of clouding the mind than antipathy is.

In the end, I'm much more interested in whether arguments are good or bad
than in the personal reasons behind the errors in the bad ones.  To tell
whether an argument is good or bad, you don't need to worry about what may
have unduly influenced someone's thoughts; you simply examine the
argument.

Regards,

Doug Axe
```

Figure 4.1 Barry Palevitz, a University of Georgia biologist and a contributing editor to a monthly magazine called *The Scientist,* epitomized the whistle-blowing mentality with this e-mail, which came just after I left the CPE. Because I knew anything I said in reply could appear in *The Scientist,* embedded in "pro-science" commentary from Palevitz, his reference to my reputation being besmirched had the sound of a school-yard bully taunting his next victim: *What's the matter, ya little wimp? Afraid yer gonna get a black eye?*

CHAPTER 5

A DOSE OF COMMON SCIENCE

Despite the opposition, by 2004 I was confident that I had confirmed Michael Denton's hunch that "functional proteins could well be exceedingly rare."[1] As quoted in chapter 3, Denton reckoned that accidental processes would be incapable of finding new functional proteins if their amino-acid sequences were more rare than about one in 10^{40} (1 followed by 40 zeros). Having now completed the experiments I described to Alan Fersht and the graduate students in 2002, I was able to put a number on the actual rarity—a *startling* number. With only one good protein sequence for every 10^{74} bad ones, I had found functional proteins to be roughly 10,000,000,000,000,000,000,000,000,000,000,000–fold more rare than Denton's criterion! Unless this number was overturned somehow, a decisive blow had been dealt to the idea that proteins arose from accidental causes.

Nevertheless, my expectation that this would compel evolutionary biologists to hang "Out of Business" signs on their doors proved unrealistic. The stream of scientific consensus continued to flow in Darwin's direction throughout 2004, and it still does.

I continue to press for the change of thinking I was pressing for then, and this change is as unwelcome now as ever. Real science is nothing like the utopian version I held at the beginning of my journey. The flag of materialism I mentioned in chapter 1 still flies proudly over the academy, and people working under that banner are expected to show due respect. Any serious opposition will bring the color guard out in full force, to the sound of blowing whistles.

That much is obvious to me. The harder question is how to advance the truth in the face of this opposition. My early recognition of the need to put Darwin's theory to a rigorous technical test compelled me to devote two decades of my career to that need. I'm convinced those were years well spent, and yet I've also become convinced of this equally important complementary need: since most people will never master technical arguments, there is a desperate need for a *non*technical argument that stands on its own merits, independent of any technical work.

As an expert who has been directly involved in many of the scientific studies described in the following chapters, I know the conclusions my coworkers and I have drawn are correct, and I know why the good work of others that gets used to argue against our work doesn't support those arguments. I could try to impart this knowledge to readers of this book, I suppose, but no matter how many chapters I devote to this, nonexperts will still be nonexperts after they turn the last page.

Does this matter? I'd like to say that it doesn't, and yet I have to admit that it does. I am just one expert among many, most of whom either disagree with my conclusion or are reluctant to admit that they agree. The simple accounts of protein research I give in the coming chapters are therefore sure to be criticized by other experts, which will leave nonexperts in the position of trying to figure out which scientists to believe. Now, if Darwin was as wrong as I believe he was, his theory can't possibly be defended as clearly and convincingly as it can be refuted. I will devote a whole chapter to that point. Nevertheless even poor arguments might seem to benefit from the status of the people making them. When all is said and done, then, nonexpert observers inevitably find themselves unable to do anything better with technical debates than trying to follow them and score them. That never settles the matter, though, because being scored the winner of a debate isn't the same thing as being correct.

For me there is no debate. The scientific facts are in complete harmony with the universal design intuition. The work my colleagues and I have done on proteins has completely resolved the internal conflict—for *me*. Resolving the internal conflict for *you*, however, will require something more. What is needed isn't a simplified version of a technical argument but a demonstration that the basic argument in its purest form really *is* simple, not technical.

As I thought about how to approach this, it occurred to me that I need to begin by correcting the misconception that science is something most of us will never do.

> *Because most people will never master technical arguments, there is a desperate need for a nontechnical argument that stands on its own merits, independent of any technical work.*

All Humans Are Scientists

We tend to overlook two key facts. One is that everyone validates their design intuition through firsthand experience. The other is that this experience is *scientific* in nature. It really is. Basic science is an integral part of how we live. We are all careful observers of our world. We all make mental notes of what we observe. We all use those notes to build conceptual models of how things work. And we all continually refine these models as needed. Without doubt, this is science. I have called it *common science* to emphasize the connection to common sense.

We embark on our quest to understand the world at a tender age. Long before we walk, we have constructed simple mental models of gravity and balance. Long before we put our hands to art, we have acquired notions of color, shape, and form. Long before we speak, we have learned to classify things into categories that await the terms we eventually use to refer to them. All of these model-building activities, and many more, use innate mental ability to process *data*—the information we receive from the world by observing it. Of course, we engage in these activities so naturally that we don't think of them in technical terms. My point is that they really *are* scientific in nature, whether or not we think of them that way.

For the most part, professional scientists respect this broadly inclusive view of science. Planetary scientists speak of the sun rising and setting just as the rest of us do. Why? Because those terms represent our common experience more simply and directly than a physically correct description based on the earth's rotation. Likewise, teachers introduce the technical understanding of sunrise and sunset by connecting it in a clear way to their students' more intuitive understanding. Children are not treated as fools for thinking the sun rises in the east and sets in the west because teachers know prior understanding is crucial to the development of refined understanding. The simple model isn't *wrong* in the sense of giving false predictions but merely *incomplete* in that it offers no causal insight. Children readily grasp the more complete model when they see how their simpler model fits within it.

This tendency to view prior understanding as a foundation for refined understanding, even in cases where the new replaces the old, continues into adulthood. No teacher of Newton's laws of motion starts by telling students to abandon their prior understanding of how things move. Telling young people who have mastered swimming and cycling and skateboarding that they have no experience of motion or no valid understanding of it would be ridiculous, just as it would be ridiculous to tell students at the next stage of their physics instruction that everything they learned about Newtonian mechanics is wrong. Everyone seems to recognize that the project of refining understanding presupposes both a general respect for understanding and a humble recognition that it is never perfect or complete.

Oddly, these basic courtesies are withheld when it comes to the universal design intuition. The story of oracle soup convinced us we all have this intuition, and we now see in simple terms how common science supports it. Bricks and breakfasts are made only if someone makes them. We know of no exceptions. With that assurance, we confidently apply the same intuition to primordial soup—only to be told we're wrong.

The people who correct us make no serious attempt to refine the design intuition in order to explain why it would work for one soup but not the other. We're simply expected to ignore the discrepancy. Apparently, our otherwise trustworthy design intuition must be overruled for the sake of Darwin's theory.

But intuitions aren't easily overruled. The psychology professors I quoted in chapter 2, Alison Gopnik and Deborah Kelemen, are acutely aware of this. Their proposed solution is for teachers to begin replacing their students' design intuition with the counterintuitive evolutionary alternative at an early age. As Gopnik put it, "The secret may be to reach children with the right theory before the wrong one is too firmly in place."[2] But if the design intuition is a product of common science, then surely to oppose it in the name of science is to make a big mistake.

Open Science

The realization that everyone proves qualified to do science by actually *doing* science is good news in multiple respects. First, this open view of science dispels the elitist myth I accepted as part of my utopian view of science. We can let that myth go without denying the existence of exceptional talent. The point is that even the most gifted people are still *people*—prone to all the internal tensions and contradictions that affect all humans. None of us rises above these common imperfections. Max Perutz didn't, and neither does anyone else.

Next, open science brings an end to authoritarian science by emphasizing the scientific value of public opinion. Because everyone practices common science, public reception of scientific claims is arguably the most significant form of peer review. For professional scientists to assume that public skepticism toward their ideas can only be caused by public ignorance is just plain arrogant. If ignorance is the cause, clearer teaching should be the remedy. When that proves elusive or ineffective, professional scientists need to be willing to find fault with their *ideas,* not the public.

This leads to the third piece of good news: Embracing open science empowers people who will never earn Ph.D.s to become full participants in the scientific debates that matter to them. Instead of merely following expert debates, nonexperts should expect important issues that touch their lives to be framed in terms of common science. Once they are, everyone becomes qualified to *enter* the debate. This doesn't apply to intrinsically technical subjects, of course, but the matters of deepest importance to how we live are never intrinsically technical.

TRUTH, PLAIN AND SIMPLE

According to the universal design intuition, tasks that we would need knowledge to accomplish can be accomplished only by someone who has that knowledge. The observation in the previous chapter that "making good enzymes will require a whole new level of insight"[3] seems to fit that intuition. Good enzymes come only from insight, and whatever the ingredients of primordial soup might be, insight isn't one of them. The results my colleagues and I have found over many years of working with enzymes also agree with the design intuition. When we examine the proposed ways in which accidental evolutionary processes are supposed to have invented enzymes *without* insight, we consistently find these proposals to be implausible.

The key to finding a nontechnical path to this same conclusion, I think, is to step back from the experiments that keep showing the implausibility of evolutionary scenarios and ask if there could be a simple reason *why* this is always so. Surely our immediate sense that instructions can't just surface by accident in alphabet soup is based on some simple, sound principle. And surely this same principle, whatever it is, must also explain why the remarkable proteins we call enzymes can't happen by accident. The universal design intuition stated in chapter 2 is a *law*, of sorts, that describes what is impossible, so there must be a simple explanation for why this law holds. The question, then, is *why* are tasks that we would need knowledge to accomplish never accomplished without knowledge?

The answer to this will become clear over the next four chapters, and as expected, common science will be the source. The key point to carry with us is that we shouldn't shy away from affirming the universal design intuition just because it contradicts the scientific consensus. The community of professional scientists is a reliable source for uncontroversial facts, but as we have seen and will continue to see, this community has a habit of stepping well outside that boundary—or, at least, scientists claiming the authority of this community do. Keep that in mind, and remember:

> *People who lack formal scientific credentials are nonetheless qualified to speak with authority on matters of common science.*

CHAPTER 6

LIFE IS GOOD

Having established that we're all capable of thinking like scientists and that we can't blindly accept how professional scientists think about life, our next step is to think about how *we* think about life.

"Wow factor" explains some of life's appeal, particularly in its more exotic forms, but what makes life uniquely attractive to us must be deeper than "wow." I believe it's something closer to *purpose*. Tornadoes rank high on the wow index because of their enormous power. But while tornadoes do what they do with great intensity, they don't *try* to do what they do. Spiders, on the other hand, *try* to catch insects, even as those insects try to escape from their captors' webs. The fear that tornadoes evoke in us is as real as the danger they pose, but the fear of a crouching cougar is palpably different in that it's a fear of harmful *intent*. There are no mind games to be played with a tornado because a tornado has no mind. Cougars are another matter.

Whether the actions of much simpler forms of life, such as the strange morphing of the foraging amoeba, involve awareness at some rudimentary level is anyone's guess. I suspect an amoeba is more like a machine than a cougar in that respect, though possibly very unlike a machine in others. There is at least a superficial resemblance between certain machines and simple forms of life. For example, if I had to pick a kind of machine that resembles amoebas, it would be those creeping robots used to clean swimming pools. They forage endlessly for debris instead of food, but their movements are almost lifelike in their complexity.

Pool robots should convince us that a thing need not be conscious in order for us to perceive intent when we observe it. Anyone watching from poolside would notice that the little details of the robot's momentary behavior—crawling along the pool bottom, climbing the wall, audibly sucking up water and debris as it reaches the surface, turning back into the water—add up in a coherent way to a higher level of behavior that we associate with purpose. Someone seeing a pool robot in action for the first time would piece these observations together after watching for a few minutes: "*Aha!* That gizmo is cleaning the pool!"

The same goes for life—only more so. A child's first experience watching a spider building a web brings particular excitement at the point when all the little busy movements are seen to add up to a whole design that is visually striking. In this the child recognizes intent even if the function of the web remains mysterious. And so do we. The whole result takes on a conceptual significance that rises above anything we perceive in the small momentary actions themselves—the bending of a leg joint, the grasping or releasing of a fiber, the starting or stopping of silk extrusion. We can easily imagine a succession of similar small actions adding up to nothing but a mess, but what we see instead is nothing of the kind. The little actions turn out to be significant because they produce a significant *end,* and we can't avoid the conviction that this was the *intended* end. The busy little spider was busy for a reason.

Activity doesn't always produce that conviction. Sometimes the total effect is just a simple sum of the momentary effects. A little rain on the street produces small puddles, then bigger ones if it continues. But even if the rain continues until the street is flooded beyond use, we aren't left with the impression that the rain or the clouds *intended* to close the street. Rain gives no appearance of being clever, no appearance of having imagined something and then labored skillfully to bring it about. In sharp contrast to the work of the spider, heavy rain has no greater conceptual significance than light rain. Much rain may bring with it important practical consequences, but we need no new concepts to understand this. If you know what *rain* and *much* mean, then you know what *much rain* means.

Rain happens, but life *doesn't* just happen—or at least this is what I hope to convince you of. Life is so different from rain that we will need new vocabulary even to think about it clearly.

Busy Wholes and Whole Projects

According to the Oxford dictionary, a *whole* is "a thing that is complete in itself." Spiders and pool robots are wholes in this sense, whereas piles of sand and thunderstorms are not. Conditions that shorten a thunderstorm or actions that divide a sand pile leave us with things that are comparable to the original, though smaller. By contrast, dissection of a spider or disassembly of a pool robot leaves us with remnants or pieces—things that aren't at all comparable to the wholes from which they came.

The same can be said of a carbon atom or of the sun—both have characteristics that don't come from a simple sum of their parts—yet neither of these objects manifests *intent* the way a spider and a pool robot do. We therefore have in the spider and the pool robot examples of a special kind of whole—the kind that manifests intent by undertaking and completing a project. Before developing this idea further, I should say that the existence of this special class of wholes doesn't in any way imply that the things that lie outside it, things like atoms and stars, are unintended or unremarkable. My point is simply that none of those excluded things *labors* the way a spider or pool robot do, as though they have *their own* intent.

We need a term to describe these special wholes—the ones that do look as though they're trying to accomplish something. As a simple way of conveying the underlying idea, I'll refer to such things as *busy wholes.* A busy whole, then, is an active thing that causes us to perceive intent because it accomplishes a big result by bringing many small things or circumstances together in just the right way. The big result is also a whole, which we will call a *whole project.* So busy wholes are whole things that tackle whole projects. When we see a finished whole project and recognize it as such, we automatically perceive intent, whether or not we saw how it was accomplished.

TWO TERMS WE WILL BE USING

whole project:

a big result accomplished only by bringing many small things or circumstances together in just the right way

busy whole:

an active thing that accomplishes a whole project

Our design intuition offers a clear interpretation of this perception. On recognizing a situation or an object to be a finished whole project, we realize that work was required to bring it to completion. More specifically, we realize that *skilled* work was required—work that brought the right things together in the right way. In our experience, skill always requires *discernment*—the ability to distinguish the right things from the wrong things and the right way from the wrong way—and discernment in turn requires *knowledge*. The moment we recognize this—that a project that requires knowledge has been completed—we immediately infer that one or more *knowers* must have been behind the work. This follows naturally from our design intuition.

Notice that this reasoning moves from the result—the completed project—to the active thing that did the work. Also notice that knowledge and intent are inferred in a way that doesn't require us to know *who* knew or intended. When we watch a pool robot do its work, we see that all its little actions add up to a completed whole project: the cleaning of a pool. We know that tackling such projects requires knowledge, and our design intuition tells us there's no substitute for knowledge. But we don't for a moment think the busy whole that did the work—the pool robot—*knows* anything. Instead, we recognize that the robot is the successful outcome of a much more impressive whole project, namely the design and manufacture of a working pool robot. The scores of busy wholes who contributed to *that* project were human beings: inventors, engineers, designers, machinists, assembly-line workers, project managers, and so on. So the knowledge and intent we perceive when we observe the pool robot in operation is ultimately traceable to human knowledge and intent, though the perception occurs whether or not we do the tracing.

Busy wholes tackle their projects by breaking them down into smaller projects in an organized way. Big projects are divided into smaller subprojects, which may themselves require further division. As long as the subprojects are complex enough to qualify as whole projects, we perceive their accomplishment to be driven by intent, and whatever does the accomplishing is therefore a busy whole. In other words, large busy wholes tend to have layers of smaller busy wholes within them, each dedicated to tackling its own subproject.

These ideas are more familiar than they sound. For example, winning a tennis match may be a whole project for a tennis player. Her success, however, is critically dependent on too many subprojects for us to list. One is the transfer of oxygen and carbon dioxide to and from her blood, which is a whole project in itself. The busy whole undertaking the top-level project (playing tennis) is a human being, whereas the busy wholes undertaking the major physiological processes supporting that top-level project are systems and organs within her body. She has a pair of lungs, busy wholes hard at work tackling the breathing subproject, hopefully well enough that she can focus on tennis. Working along with her lungs on the breathing subproject are other busy wholes, such as her nervous system and her diaphragm. And as you might expect, the projects being accomplished by each of these anatomical busy wholes may be further subdivided into projects assigned to tissues, and then to cells, and then to subcellular structures and intercellular interactions, and finally down to the molecules of life.

Now, the question that most interests us is whether anyone intended for our lungs and the cells within them to tackle their respective projects the way the tennis player intends to win her match. Are we right to infer purposeful design when we watch the human body—or any living body—in action, the way we infer it when we watch the pool robot? The answer to this will emerge as we continue, but the points to grasp here are more modest. First, rightly or wrongly, we're naturally inclined to think that things like organs and cells *were* intended, and second, a commonsense rationale can be offered for this inclination. Again, whether the rationale I've offered is the one we actually use isn't our concern. Rather, our interest is to decide whether the inclinations themselves are correct.

As our journey continues, I will build a case for thinking they *are* correct, but my objective is as much to inspire as it is to convince. If Darwin's theory has left us with an impoverished view of life—as I believe it has—then there is as much to be gained by articulating a more satisfying view of life as there is by showing that Darwin was wrong. I hope to do both.

Of Salmon and Orca

I have come to think that everything about a salmon is *salmon* and everything about an orca is *orca*. Having worked in molecular biology for decades, I know the similarities between these two aquatic animals are real and significant, but I confess that this head knowledge vanishes when I watch mature salmon, having spent most of their lives in the salty waters of the Pacific, fighting their way upstream through freshwater to reach the place where their lives began. Their mission literally consumes them. Forsaking all food, they sacrifice every ounce of their flesh, launching themselves over and into rocks as they battle their way up rapids—all for their final purpose of parenting offspring they won't live to see.

The salmon's way of passing the baton from one generation to the next may look brutal to us, but that concern doesn't seem to have crossed their minds. Nor has it crossed mine on the occasions when I've watched them. In their uncompromising determination, these magnificent creatures make it abundantly clear that they're doing exactly what they were meant to do—like heroes and heroines rushing into their last battle.

Most of them perish in earlier battles. I spent a day watching this too, with a small group of friends on a whale-watching boat in the Strait of Juan de Fuca. Orcas, often called killer whales, spend their long lives in family groups called pods. They're the most formidable hunters of the ocean, fearing nothing and feeding on whatever looks good to them, including the otherwise invincible great white shark. Like sharks, orcas kill, but the *way* orcas kill is altogether different. They are clever and graceful, as greatly to be admired by us as they are to be feared by fish.

The captain of our vessel, an expert orca watcher, located a large school of salmon with his sonar equipment. With the engine off, we sat above the school for several minutes and watched the whales do their thing . . . with style. It quickly became evident to me that these creatures are smart enough to know they're being watched and gregarious enough to seize any opportunity to show off. As though executing a play from their hunting playbook, they confined the salmon by using a corralling technique where pod members take up positions around the perimeter of the school to prevent it from dispersing. From our above-water vantage point the signs of this were occasional spouting at some distance from the boat in all directions. The salmon, of course, had a much better view of what was happening, though presumably not the favorable impression of it that we had.

What we saw from the deck of the boat was unforgettable. These elegant show-offs took turns swimming at high speed through the trapped school of salmon, gobbling one or two with each pass and celebrating their success with breathtaking high breaches—five tons of slick black and white launching out of the water with implausible ease. Gravity was repealed for a moment as they took to the air. In the space of a breath—the half second I needed to untangle seeing from believing—*flight* actually crossed my mind, only to be dispelled by the thunderous crash of reentry. How such extremes of mass and grace can possibly reside in the same skin remains a mystery to me.

The thought that this brief spectacle meant quite a few salmon would never make their heroic end-of-life journey only occurred to me later. When it did, the salmon saga again seemed more like valor than tragedy—not because salmon can be virtuous but because there is something intrinsically beautiful about what they are and something magnificent about the intensity with which they live that out. What emerges from their heroism is deeply compelling.

Orcas are equally compelling in their own distinctive way. And somehow—for me, anyway—the fact that part of what it means to be a salmon is to sacrifice yourself or some of your relatives to feed the orcas makes neither species less magnificent or less compelling.

Life à la Darwin

If there's anything compelling about Darwin's view of life, it's the simplicity of his core idea. Underlying the jumble of ideas that evolutionary biology has become is one crisp principle: things with the ability to reproduce themselves automatically carry the potential to produce descendants that are *better* reproducers. Few theorized explanations are so disarmingly direct. The perpetual improvement of reproducers *seems* to require only (1) that they carry out their reproduction imperfectly—with small errors (mutations) being introduced occasionally—and (2) that at least some of these errors enhance reproduction, if only slightly.

The problems with this idea will become clear later. For now, I simply ask how this explanation, if we accept it, ought to shape our picture of life. Considering how innocuous the assumptions appear, the depth of their implications may come as a surprise. We are left to view the many kinds of life much the way we view geological features: as things in constant flux. Mountains seem permanent to us because they retain their shapes for long periods, yet we know they're constantly being reshaped by the natural forces that formed them. The same must be true of life if, like the earth's crust, it is constantly being molded by unseen forces.

By this view, there must have been a simple ancestor of all animals, whose offspring were pushed by natural selection in many different directions, like leaves dispersed over water by convective currents. Modern animals must then be nothing more than the present locations of those drifting leaves. Each is like one frame in a long time-lapse video—the snapshot of the day. The magnitude of the cumulative changes may amaze us as we contemplate the staggering variety of animals that came from that single ancestor, but nothing about the present forms *themselves* should amaze us, as this would be like being amazed by one frame in the middle of a long video. Presumably the descendants of today's spiders and whales and salmon will become as radically different again, given as much time again.

Though I personally dislike this fluid view of life, I would have to come to terms with it if I were committed to the idea of natural selection being life's creator. That wouldn't be easy. I would be constantly bothered by the contradiction between this view and what I see when I open my eyes, because life looks profoundly unlike geology to me. The things of geology are best understood by grouping them into a relatively small number of categories, whereas biology calls for a different approach. Serious pursuit of a satisfying understanding of life's distinct varieties forces us to abandon the idea that they're all fundamentally the same thing: reproducers stumbling along toward better reproduction. The spider, the salmon, and the orca will have none of that idea. Each is strikingly compelling and complete, utterly committed to being what it is. Each will finish heroically by death or even by extinction, but not by surrendering to forces that would turn it into something else.

Perfection and Its Critics

This theme of commitment takes the idea of wholes to a new level by hinting at the possibility that some wholes are what they are because they *ought* to be so—as if they are expressions of something truer and more significant than any temporary physical representation. The idea here is not that some things are so good that they had to exist but rather that some things are so good that they cannot be other than what they are. The tapestry of human creativity is adorned with several examples: a perfect musical composition, a perfect poem, a perfect mathematical proof—timeless treasures to be beheld but never to be reworked.

Life is the quintessential representation of this idea, utterly without rival among human works. Forget the old textbook definition of life—something to the effect of life being a self-perpetuating, nonequilibrium process based on carbon chemistry and driven by the influx of solar energy. That never resonated with anyone who mused on life. No, life must be something much richer, immeasurably more worthy of our attention. Life is mystery and masterpiece—an overflowing abundance of perfect compositions. You and I are among them, here for a brief time to delight in as many more as we possibly can.

Surely everyone senses the profound wonder of life. It seems too overwhelming to be overlooked. Equally obvious is the tension between this sublime view of life and the explanation offered by Darwin. His idea that life wanders from one variation to the next, never committing, always yielding to the blind force of natural selection, is plainly incompatible with the idea that the physical forms of life are expressions of something deeper, something immovable, something perfect.

Darwin's idea that life wanders from one variation to the next, never committing, always yielding to the blind force of natural selection, is plainly incompatible with the idea that the physical forms of life are expressions of something deeper, something immovable, something perfect.

So how might a person who's reluctant to abandon Darwinism respond to this high view of life? I've seen two approaches. The cruder of the two, and probably the better known, is to downgrade the high view. Of the various downgrade paths, one emphasizes aspects of life that we all agree are not right. Birth defects, cancer, infectious diseases, parasites, suffering, and loss of biodiversity are all disturbingly bad, so it's reasonable to offer these as evidence against any claim that life as we see it is comprehensively good. But my point is more subtle than that. I'm not denying that the present state of life is troubling in many respects. Rather, I'm affirming that something spectacularly *good* is clearly discernible even through the haze of trouble.

Another way of downgrading life is to assume the role of a bio-critic—someone who looks for faults in the design of living things. As one example, the giant panda has a protruding bone in its wrist that serves a thumb-like role, enabling the bear to grasp bamboo (Figure 6.1). The fact that this bone (called a *radial sesamoid*) isn't a true jointed thumb like ours has led some people to view it as a makeshift adaptation that no good designer would employ. Not surprisingly, others argue that it *is* a good design. For my part, I find myself evaluating the people more than the panda. None of these people, however earnest they may be, have any deep grasp of the principles of design and development underlying sesamoid bones or thumbs, to say nothing of pandas. Indeed, none of us do. Search the world's top research centers and you'll find no skeletal engineers—no one who has the faintest idea how to encase earthworms in exoskeletons or how to endow leeches with backbones. Surely, then, our total inability to answer these *how* questions categorically disqualifies us from serious engagement of the higher *why* questions. We're free to form opinions on these matters, but they're nothing more than that. My opinion, for those interested, is that the giant panda is yet another example of something perfect—something that is exactly as it should be.

Figure 6.1 Location and use of the radial sesamoid bone in the forepaw of the giant panda.

The better option for people who aren't ready to part with Darwin's theory is to embrace life's excellence in the hope that this will ultimately prove explicable in Darwinian terms. This option has the considerable advantage of affirming our high view of life, but with that comes the challenge of making a square peg fit a round hole. If natural selection is not just the master shaper but also the incessant fiddler, as Darwin thought, then evolution never reaches a compelling end. In his own words,

> It may be said that natural selection is daily and hourly scrutinising, throughout the world, the slightest variations; rejecting those that are bad, preserving and adding up all that are good; silently and insensibly working, whenever and wherever opportunity offers, at the improvement of each organic being in relation to its organic and inorganic conditions of life.[1]

For Darwin, then, the thought of all the various evolutionary lines terminating at ends that are too good to be altered would have been as inconceivable as the thought of unchanging conditions. Local ecosystems and climates experience one change after another, which means conditions never settle into a permanent state, which means the work of natural selection is never finished. By contrast, the affirmation that there is something uniquely compelling about living things as we now see them is an affirmation of completion. It rejects the idea that the designs of life are like leaves drifting on a pond, or like ever-changing mountains, or like frames in a video. So followers of Darwin seem to be faced with the dilemma of deciding whether to believe their theory or their eyes.

To understand this dilemma more clearly, try to imagine a plausible evolutionary precursor of all modern animals. Having descended from the very simplest life, this creature would have possessed only the most basic characters found in all modern animals, most notably a multicellular body that differs from plants by lacking cell walls and photosynthesis. Among modern animals, *sponges* come closest to meeting this description. So picture an ancient creature of sponge-like simplicity.

Now, if this ancient sponge really produced the modern orca through a long succession of intermediates, we should ask: What *drove* the astounding transformation of animal form along this particular line of descent? There seem to be only two possible evolutionary answers. Either the "conditions of life" determined the form, or natural selection did. That is, either selection delivers whatever the conditions of the day call for, or selection steers its own course to a highly fit end, dealing with the changing conditions along the way. The first implies that ancient sponge and modern orca are connected by a succession of comparably fit animal forms, whereas the second implies an upward progression from the inferior ancient form to the superior modern one.

Both scenarios have issues. If we say conditions are in the driver's seat, we're saying life is noncommittal to the point of incoherence—open to being either a sponge or an orca or any of the subtle gradations supposed to span that not-so-subtle gap. On the other hand, if we say selection is in control, then we come uncomfortably close to *personifying* evolution, as though it had both the vision to know what it wanted to do with that crude sponge and the patience to walk it through a long period of awkward adolescence, knowing how good the end result would be.

WHY PROTEINS DON'T EVOLVE (ANYMORE)

The bigger question, though, is whether life is open to evolutionary reshaping *at all*. The answer that has emerged with increasing clarity in recent years would have surprised Darwin.

Some of the key facts take us back to the subject of proteins. To explain how natural proteins, with their exquisite functions, could have appeared by accident is a monumental challenge. This challenge can be divided into a more extreme aspect and a less extreme aspect, both of which are proving to be major obstacles for evolutionary theory. The more extreme challenge is to explain how mutations and selection could have produced completely new structural themes for proteins, called *folds* (Figure 6.2). The less extreme challenge is to explain how mutations and selection could have produced functional variations on existing fold themes.

My colleagues and I have studied both of these challenges. To focus on the less extreme one, biologist Ann Gauger and I chose to work with two strikingly similar yet functionally distinct natural enzymes, which we'll call enzyme *A* and enzyme *B* (Figure 6.3). Our aim was to determine whether it would be possible for enzyme *A* to evolve the function of enzyme *B* within a time frame of billions of years. If natural selection really coaxed sponges into becoming orcas in less time, inventing many new proteins along the way, we figured it should have ample power for this small transformation. But after carefully testing the mutations most likely to cause this functional change, we concluded it probably isn't feasible by Darwinian evolution.[2] Additional work supports this conclusion. Mariclair Reeves—like Ann Gauger, a biologist at Biologic Institute—painstakingly

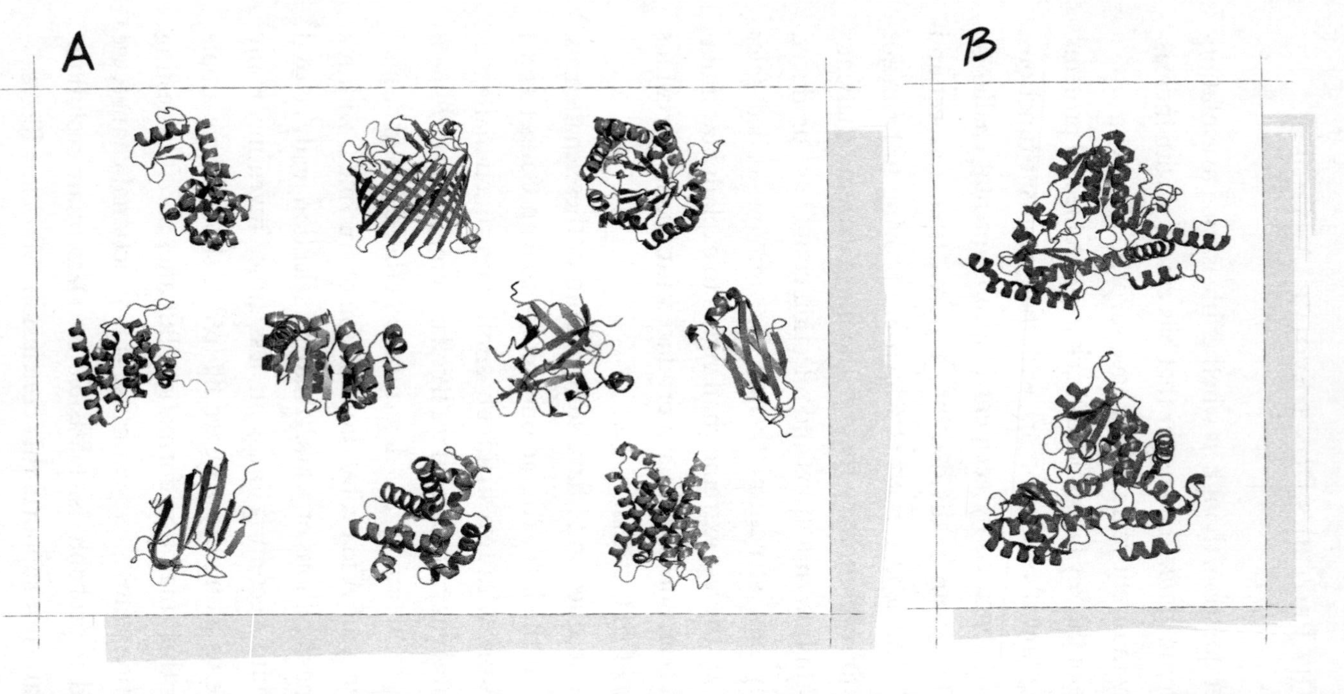

Figure 6.2 The distinction between structural themes and variations on a theme for proteins. Of thousands of known structural themes, or *folds,* ten are shown as ribbon diagrams on the left (A). Notice the great variety of folds that can be made from the two basic structural elements, the alpha helix and the beta strand. Figure 6.2B shows two variations on the same fold theme. Although the boundary separating variations on a theme from differences of theme is imprecise, these two categories have proved useful for classifying protein structures.

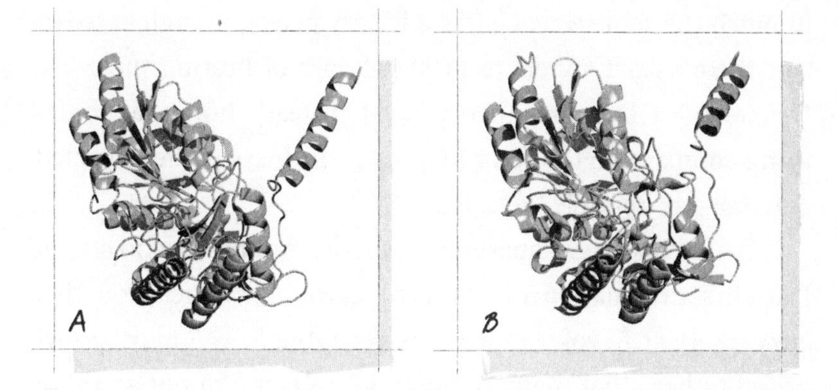

Figure 6.3 Both enzymes that Ann Gauger and I chose to study are formed by two identical protein molecules that grip each other in a manner analogous to a handshake. Here you see the striking similarity between the "hands" that form enzyme *A* and enzyme *B* (their actual names are Kbl and BioF, respectively).

tested millions upon millions of *random* mutations, searching for any evolutionary possibility that we may have overlooked in our first study. She found none.[3]

We've received two good questions about this result from nonscientists. The first is how it's even possible to test a process that takes so long. Certainly we can't observe anything over eons, but you're familiar with the possibility of calculating times for processes that are too slow to watch. To estimate how many years it will take a young tree to reach a desired size, we measure the growth over one year and then compare this to the additional growth needed. Estimates for processes involving *chance* carry a slight twist. If you know the proportion of lottery entries that win each week, for example, then you know how many times players should expect to have to enter before they win. Individual players will do better or worse than expected, but the average long-term result should be as expected. Scientific theories that involve chance, as Darwin's theory does, are analyzed in much the same way. Unlike a lottery player, though, a scientific theory can't appeal to luck in hopes of beating the odds. Defenders of Darwin's theory must instead show that something comparable to life in its present fullness is the expected outcome once simple life exists.

The second good question is whether scientists who accept Darwin's explanation of life also accept our conclusion that enzyme *A* can't evolve to work like enzyme *B*. You may be surprised to hear that many of them do. In fact, I'm not aware of anyone having challenged that conclusion. How, you may wonder, can anyone believe that natural selection is incapable of such a tiny transformation while maintaining that it accomplished so many gargantuan ones? The current answer from evolutionists is that natural selection is a victim of its own success. That is, natural selection is now thought to have been so effective at tailoring organisms to their environments that it *did* reach end points—creatures so good at being what they are that they can no longer undergo evolutionary change.

Berkeley paleontologist Charles Marshall exemplifies this perspective in his critical review[4] of Stephen Meyer's book *Darwin's Doubt*.[5] Meyer's thesis is that Darwin's evolutionary mechanism is incapable of generating new animal forms, in part because it's incapable of generating new protein forms. In reply, Marshall suggests that new animal forms evolved without any need of new proteins.[6] According to him and others, this was done by "rewiring" the gene regulatory networks (GRNs) that control when and where existing genes are turned on within a developing embryo. Marshall concedes that experimental alterations of these networks usually kill developing embryos, but he thinks this should be overlooked because "today's GRNs have been overlain with half a billion years of evolutionary innovation (which accounts for their resistance to modification), whereas GRNs at the time of the emergence of the phyla [the basic animal forms] were not so encumbered."

Marshall and I agree, then, that life in its present form resists evolutionary change. We disagree on the likelihood of natural selection ever having done anything remarkable, but we both end up favoring an explanation of life that looks more purposive than Darwin's explanation. If natural selection really shaped life, it worked more like an artist shaping clay than erosion shaping sandstone. It was skillful enough to transform the ordinary into the extraordinary, and wise enough to know when that work was finished.

The molecular version of that view has become the main criticism of the conclusion I reached with Ann Gauger and Mariclair Reeves. We were wrong, critics say, to expect enzyme *A* to be capable of further evolution because enzymes, like animals, have been perfected to the point where they're no longer pliable in the hands of natural selection. Dan Tawfik of the Weizmann Institute, whom I know from my days in Cambridge, is a champion of this idea. Believing that "broad-specificity enzymes served as progenitors for today's specialist enzymes,"[7] Tawfik would presumably agree with the critics that Ann and I were wrong to expect today's specialist enzymes to evolve the way yesterday's broad-specificity ones did.

Whether this latest version of evolutionary thinking is any more plausible than previous versions will become clear as we proceed. To his credit, Tawfik recognizes the difficulty of explaining how the supposed broad-specificity enzymes would have arisen in the first place. Since they had to be true enzymes—

folded proteins with geometrically complex active sites—it's unclear what's been gained by proposing them as precursors. His own diagnosis of this is admirably frank: "Evolution has this catch-22: Nothing evolves unless it already exists."[8] In other words, don't expect a working *X* (you name it) to come out of the evolutionary process unless a working *X* went in.

Again I find myself in agreement, and that makes resolution of the dispute among scientists seem hopeful. The most sensible question to ask next, though, will put this hope to the test: *What's left of a theory of origins once it has been conceded that it doesn't explain how things originate?*

We've seen in this chapter that living things are exquisite wholes—so committed to being what they are that they give the distinct impression of being *meant* to be what they are. With that realization in mind, we're ready to begin scrutinizing the opposite view. If life was *not* meant to be, then it is accidental, and of the very few suggestions for how it *could* be accidental, none has had more hopes pinned on it than natural selection. Accordingly, we will use the next chapter to examine natural selection under the powerful lens of common science.

CHAPTER 7

WAITING FOR WONDERS

All attempts to explain how Earth came to be teeming with life must face the challenge of explaining extraordinary things. Ordinary physical causes seem adequate for explaining things that aren't task-oriented (things like atoms and stars and tornadoes), but our design intuition tells us those causes can't explain the things we're calling busy wholes (things like spiders and pool robots). Many scientists tell us otherwise, that ultimately *everything* is rooted in ordinary physical processes. Those processes, they say, turned primordial soup into simple life, and simple life into simple animals, and simple animals into complex animals, some of whom make robots. If those scientists are right, extraordinary things don't really require extraordinary causes after all.

This doesn't sit well with our design intuition, though. When we see working things that came about only by bringing many parts together in the right way, we find it impossible not to ascribe these inventions to purposeful action, and this pits our intuition against the evolutionary account. As our rejection of oracle soup showed, people differ not in whether they have the design intuition but in whether they have exempted evolution from it. We all agree that a spider's web is the product of the spider's spinning instinct. The point in dispute is whether anyone *intended* for spiders to have that instinct, or the body parts that enable it to be used. If no one meant for spiders to spin, then spiders were invented by accident, making our design intuition deceptive. If someone did, then spiders were deliberately invented, making the evolutionary account deceptive.

The way forward is to recognize that whatever value we place on the design intuition, we can certainly reason without it. Without rejecting intuition, we can lean instead on observation and calculation to decide whether we should expect a universe like ours to produce busy wholes like spiders. The important question, then, is whether evolutionary theory is more in touch with our observations than our design intuition is.

Is it? What fact did Darwin cite that should, contrary to our intuition, make us expect things like sponges to be converted into things like orcas? What cause did he identify that has the power to make such extraordinary transformations so easy they happened a million times over in different ways? What could possibly tame such frightening improbabilities? The standard answer even today is natural selection—the tendency of more fit organisms to have more offspring. No one disputes this tendency, but can it really work these wonders?

Robotic Football Fans

A thought experiment will help point us to the answer. The one I have in mind has to do with football, of all things. As everyone who's been to a professional football game knows, football crowds are loud. This is true in the United States, where footballs are oblong, and equally true in the rest of the world, where they're round. Fans of my local team, the Seattle Seahawks, have taken the phenomenon to what some would call an unhealthy extreme. Exactly two months before the Seahawks won Super Bowl XLVIII, their fans earned the world record for the loudest crowd at an outdoor sporting event, reaching an ear-splitting 137.6 decibels on the second day of December in 2013.[1]

Because football crowds are characteristically loud, then, might something with absolutely no understanding be able to find its way to a football game just by seeking loudness? I'm thinking of a noise-seeking robot. Imagine a weatherproof robot that can be dropped by parachute to any location, land or sea. Upon landing, it detaches its parachute and begins homing in on sources of sound. First it uses a directional microphone to measure sound in all directions from its present location. Then it swims or crawls a short distance in the direction of the loudest measured sound, after which it stops to repeat the measurement. This listen-and-move homing cycle is repeated long enough for the small moves to add up to a considerable distance, though how far this takes the robot will depend on how straight the course is.

The question is, what would have to happen for a robot of this kind to find its way to a roaring football stadium? Being dropped within earshot of the crowd's roar, or at least in close proximity of earshot, would be helpful. Even then all kinds of things could go wrong. Competing sounds—like that of traffic on the streets—could interfere. The football crowd would probably generate much more noise than the traffic, but because sound falls off with distance, nearby traffic may register as louder than distant cheering. Still, there is at least a small hope that the robot would find its way to a game if it were dropped within earshot of a stadium.

But suppose the drop point is entirely *random*—anywhere on Earth with equal likelihood. In this case, the odds of success would be very slim indeed. Even when we take all the world's football stadiums into account, the earth is so large in comparison to the regions within earshot of stadiums that there's little chance of our robot hearing even the faintest sound of a football crowd. As powerful as the crowd's roar is inside any stadium, it's completely inaudible over most of the planet. Our poor robot would probably end up on a shoreline somewhere,

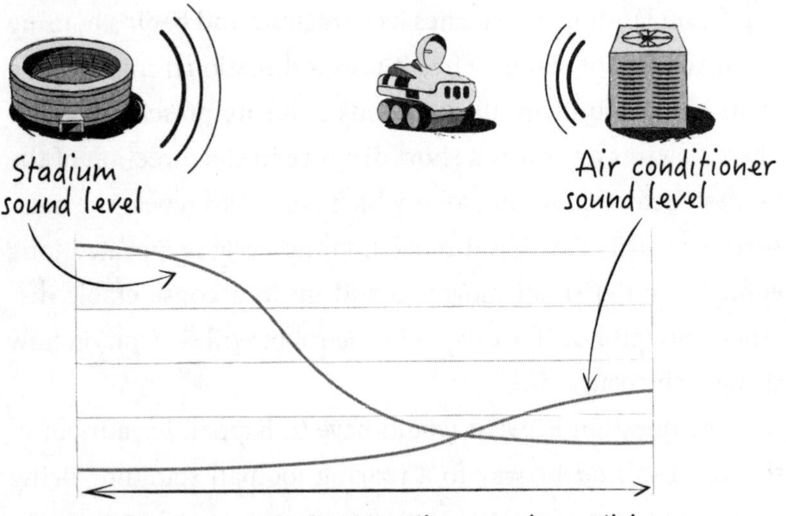

Figure 7.1 The noise-seeking robot's dilemma. Even in locations where the roar of the stadium crowd is audible, there are competing sound sources. Although no single competing source is producing as much sound as the crowd, the robot will register nearby sounds as louder than distant sounds. Combined with the abundance of competing sound sources, this makes homing much less reliable than it would otherwise be.

enticed by the sound of crashing waves. Even if it were to land in a city that has a football stadium, being drawn to an air conditioner or getting hit by a bus would be more likely than making it to a football game (Figure 7.1).

We can get a more accurate assessment by looking at some numbers. If we assume there are about two thousand major football stadiums in the world, each of which can be heard up to about a kilometer away (about two-thirds of a mile), then all of these within-earshot areas amount to about six thousand square kilometers (about two thousand square miles). This is a meager thousandth of a percent of the earth's half billion square kilometers of surface, which means the probability of our robot landing at a spot where a football crowd can be heard—if the timing is right—is a mere one in a hundred thousand.

Nevertheless the robot's movements might bring it within earshot of a stadium eventually. Success is unlikely in the short term, but if we suppose our robot runs on solar power and is built to last, the odds should increase with time. Unhelpful noise sources will almost certainly be homed in on for a long time, but eventually changes of circumstance will cause the robot to leave each of these distractions and move on to something new. This could happen in any number of ways. Maybe the force of a wave or the paw of a curious bear will push the robot to a location where new sounds are audible. Or perhaps the sound of thunder or of wind in the trees will distract the robot momentarily, just long enough to set it on a new course. Because occurrences like these are possible, it's only a matter of time before they happen. Our expectation, then, is that repeated changes of circumstance will *eventually* put the robot within earshot of a football stadium. It may take years or decades or even centuries, but success must come in the end if the experiment continues long enough.

There's something odd about this version of success, though. We started by asking whether the ability to seek noise might enable a robot to find a football stadium, and now we're invoking something *other* than noise-seeking to achieve success. This other factor is *repetition*—repeated opportunities for rare favorable circumstances to be stumbled upon. To the observer, this blind repetition carries so little promise that it seems like nothing more than an interminable wait. Sure, waiting is bound to work if it can be extended indefinitely. But if open-ended waiting is really an option, just how significant was the homing ability in the first place? After all, even something completely passive—like a Styrofoam packing peanut—might make its way into a football stadium if we wait long enough.

The Connection to Evolution

That same question of significance impinges on our discussion of evolution, which motivated the thought experiment in the first place. In fact, there are strong similarities between our robot and the evolution of a species, the main one being that natural selection acts very much like homing. Just as the robot moves toward the loudest noise as judged from its present location, so natural selection tends to shift the genetic makeup of a species toward the highest fitness as judged from its present members. The robot's homing causes movement through geographic space; selection's homing causes movement through an abstract space, namely the genetic space consisting of all possible genome sequences. Each movement in this genetic space consists of a change in the genome sequence that typifies the species, taking many generations to complete.

Figure 7.2 shows what one of these changes might look like if we observed a species long enough to watch it happen. The process is nothing more than a gradual replacement of the most common genetic type (unmarked beetles, in this case) caused by the presence in the population of a more fit type (double-dot beetles). In most actual cases the types wouldn't be as visibly different as they are in this hypothetical example. Furthermore, the observer wouldn't really know whether natural selection was the cause of the change, because transitions like this often happen for reasons having nothing to do with fitness. Nevertheless, when fitness *is* the cause, the process is akin to the stepwise homing of our robot, as we will see in more detail in a moment.

Another similarity between our robot and the kind of genetic movement we're interested in is that for both we have a clear understanding of success. I chose football stadiums as the robot's objective because they're highly distinctive locations—numerous and varied but always noteworthy. Those same adjectives apply

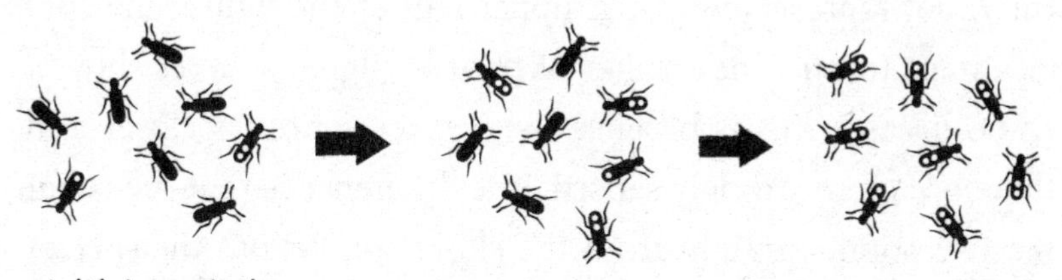

Figure 7.2 The visible work of natural selection on a hypothetical beetle species. Eight beetles represent the makeup of the population at three points in time, many generations apart. By the final point, selection has homed in on the double-dot variant, as seen by the fact that this now typifies the species.

even more profoundly to the living things that evolutionary theory must explain. We aren't asking whether natural selection causes just *any* changes. Instead, we're focusing on changes of the most noteworthy kind. We want to pinpoint anything in the evolutionary mechanism that might have the astonishing inventive power that Darwin and his followers have attributed to evolution.

Selection: Exit Stage Left

To that end, the most important thing we learned from our robot is that the mere ability to home in on signal sources isn't what brings success. Instead, success occurs when the *right kind* of source happens to be close enough to outcompete any other sources. We saw this when we recognized how unproductive homing was unless the noise being followed happened to be from a nearby stadium. If the analogy to evolution holds, we should expect something similar for the homing caused by natural selection.

As Figure 7.3 shows, the situation is indeed similar. Like the robot represented in the upper half of the figure, the species steps toward the weaker of the two signal sources shown. In both cases this is because the weaker source is closer and therefore more strongly sensed. The difference is that the robot receives sound straight from the closer source, but the species must make do with much less definitive information. All the species senses is the relative fitness of the different genomes that currently exist among its members. You might think this amounts to a huge amount of information, considering how many millions of individuals can belong to a species. However, the genetic makeup of most individuals differs only negligibly from a great many others, so relatively few genomic variations are in play at any time (represented by the dots in the lower half of the figure). The homing action of natural selection is limited to stepping from the current dot (labeled *Initial location*) to

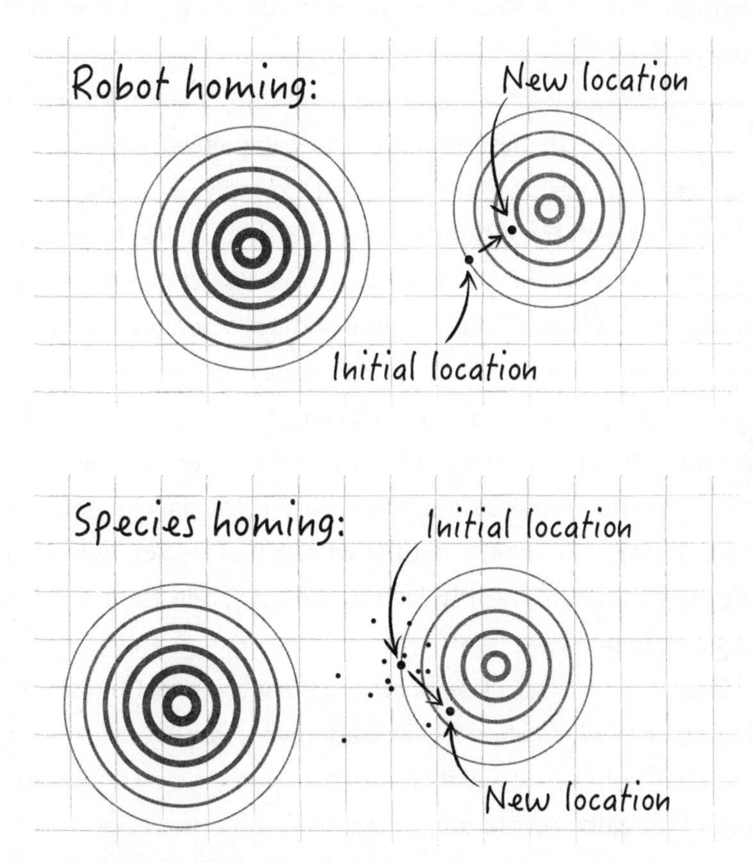

Figure 7.3 Comparing a single step of homing by our robot to a single step by natural selection. The top half is a map in the usual geographical sense, and the bottom is a map of "genetic space," meaning the space of possible genome sequences. Concentric rings mark the locations of two sound sources in the upper map and two fitness sources in the lower map, showing how the signal strength falls the farther we go from these sources. Genome sequences in the species are shown by dots. Larger dots indicate the predominant genomes before and after the homing step.

the best available dot (*New location*). The figure shows one such step. More often than not, the best available location would be the current location, so no step would be taken.

For our purposes, the crucial thing to get is that a new functional feature—an invention—produces no fitness signal at all until at least one individual in the species already has that invention—*which means natural selection itself can't invent!* Suppose, for example, that the strong source of fitness in the lower left of Figure 7.3 is a genuine invention of some kind and the weak source in the lower right is completely insignificant as far as invention goes (I'll give real examples in a moment). Oblivious to matters of significance, selection moves the species *away* from the invention because in this case the invention is completely "invisible." That is, none of the species' members has one of the special genome sequences needed for the invention to be produced. Of course, if I had placed another dot close to the invention, then the result would have been different. The point is that this dot would belong there only if the special genome sequence it marks *already existed* within the species. Selection can't place the dots. It only *follows* the dots, and then only in this shortsighted way.

So by the time selection begins to favor an invention, something *other than selection* has already invented it. This is one of those common-science gems to be treasured—an obvious realization that gains revolutionary status for no other reason than a long tradition of ignoring it. Let this soak in for a moment. Despite all the grand claims—everything from the popular plea of Richard Dawkins's *The Blind Watchmaker*[2] to the technical pitch of Graham Bell's *Selection: The Mechanism of Evolution*[3]—the very logic of natural selection assures us that the power of invention resides elsewhere. And because evolutionists have never agreed on what this elsewhere is, the gaping hole that has always existed in the middle of evolutionary theory is still there.

Dan Tawfik hit the nail on the head: Nothing evolves unless it already exists.

> **THE GAPING HOLE IN EVOLUTIONARY THEORY**
>
> *Evolutionary theory ascribes inventive power to natural selection alone. However, because selection can only home in on the fitness signal from an invention after that invention already exists, it can't actually invent.*

The response to this gem from defenders of Darwin's theory is that selection didn't produce eyes or brains or lungs all at once. Long before it worked upon complex organs like these, it was refining the many simpler things that paved the way for these greater things to appear. Each of the simpler things was beneficial for its own reason, we are told, so selection was able to work even though the grand functions didn't yet exist.

Although this response continues to put selection forward as the hero of the story, again something else has to be doing all the remarkable work of invention. Selection can cause a species to take genetic steps, but without any way of *directing* those steps, movement of that kind wouldn't go anywhere. To reach an interesting destination requires not just steps but *coordinated* steps, helped along by nicely arranged *stepping stones.*

Suppose, for example, that some biological feature—call it *X*—performs its function by means of numerous component functions. For *X* to work requires, say, a working *P* and a working *Q* and a working *R*, and for *P* to work requires a working *H* and *I* and *J* and *K*, and so on. In light of all these requirements, how could the invention of *X* have come about by accident? What is supposed to have coordinated the appearance of all these necessary things at the right times and places, laying the stepping stones to *X* out so insightfully? To say merely that the precursors for each necessary part were selected for different reasons is to ignore the uncannily complicated circumstances that would be needed for this to be so. After all, knowing that a certain species would benefit from a working *X* gives us no reason to believe that precursors to all the components needed to build *X* would just happen to have been beneficial earlier, each for its own reason, or that all these precursors could have been coaxed by small modifications into their new *X*-critical roles just when *X* was needed. Only in action films—where realism isn't even on the agenda—do things fly together in such helpful ways.

Science, on the other hand, should view claims of helpful coincidences of this kind with suspicion. At best, they're a misinterpretation of history, where selection—the sham hero—steals glory from an unnamed hero working behind the scenes to make everything come out right.

Odds Beaten . . . or Bypassed

We'll soon see why accidental invention *must* be highly improbable. In the meantime, if we provisionally accept this on the basis of our intuitions, we're prompted to wonder how an explanation of life can stand in light of these improbabilities, not having overlooked them but having properly dealt with them. There seem to be only two possibilities. Either an explanation *beats* the improbabilities (that is, it counters them with something equally potent) or it *bypasses* the improbabilities (that is, it renders them irrelevant).

As you may have noticed, the improbabilities we're discussing are easily bypassed by following our design intuition. Explanations of life that credit the invention of living things to a *knower* avoid the burden of improbability altogether. As for beating the improbabilities, the tactic we resorted to with our robot turns out to be the *only* tactic. That is, the only way to beat improbabilities is to have such an abundance of opportunities for the unlikely outcome to happen that its occurrence is no longer unlikely.

Another serving of oracle soup will help us see this. Picture a chef presenting a pot of alphabet soup and lifting the lid to reveal written instructions. Now, ask yourself what would qualify as a satisfactory explanation for what you just witnessed? Someone having spent a couple of hours in the kitchen arranging the letters would qualify, but that would be bypassing the improbability, not beating it. My question is, supposing the chef insisted the instructions were formed by nothing more than the process of boiling and cooling the soup, what could conceivably satisfy you that he or she is telling the truth?

I hope you wouldn't fall for an authoritarian approach. Imagine a team of physicists, all committed to the materialism. Would you be persuaded if they gave you a series of technical lectures claiming that the physical causes that wrote genetic instructions in primordial soup did their work again in alphabet soup? Surely not.

To stand your ground in the face of that kind of intellectual intimidation, you'd need a simple, unassailable common-sense argument, and that's exactly what you'd have. No amount of technical mumbo jumbo can change the fact that it's extremely improbable for accidental causes to do the work of insight. If the physicists attribute the instructions in the alphabet soup to "correlative entrainment"—whatever that means—your first question should be "Did this 'correlative entrainment' receive any assistance from someone who understood the instructions, or was it a completely unguided physical process?" And if the answer is that it was unguided, your next question should be "Of all the possible outcomes an unguided process might have produced, how was this 'correlative entrainment' so fortunate as to achieve such a *special* outcome—one that looks for all the world as though it *was* guided?"

There is no credible answer. Insight is absolutely unique, without rival among the mindless causes to which materialists limit themselves and, as we will later see, not reducible to those causes either. Being fundamentally unlike insight, physical causes can't do what insight does in any systematic way. Sound waves are unlike water waves in their physical substance, but the fact that they're both waves means they show strikingly parallel behavior in many respects. Parallels for *insight,* on the other hand, are nonexistent.

The lack of any parallel to insight means that any instance of mindless causes doing the work of insight would have to be a fluke . . . a coincidence. Minor examples abound. Short words do appear in alphabet soup from time to time, not by any mysterious force working in the broth but by coincidence. Indeed, our robot exercise showed that the improbability of coincidence can be offset by repetition, at least to a degree. Whether repetition can conquer the challenge of biological invention is therefore worth considering. Certainly, billions of organisms propagating through millions or billions of generations is repetition on a whopping scale. Perhaps, then, we'll find that our design intuition isn't calibrated for use on such a scale.

A CONSEQUENCE OF THE UNIQUENESS OF INSIGHT

The lack of any parallel to insight means that any instance of mindless causes doing the work of insight would have to be a coincidence.

We can imagine being surprised, anyway. Continuing our thought experiment, suppose that after you express your skepticism to the chef, he or she leads you through the swinging doors, whereupon you discover that what you thought was a kitchen is really a logistics center—the Mission Control for a massive soup-boiling operation involving a hundred million square feet of kitchen space scattered over six continents! With the aid of automated text detection, human operators are alerted whenever the pasta letters in any of millions of cooled pots resemble instructions, at least to the eye of a computer. After running this operation at full tilt for just over nine years, one of those alerts turned out to be the real thing. The contents of that winning pot were carefully frozen to keep the letters in place for transport from a kitchen facility on the outskirts of Johannesburg, and you were fortunate enough to be invited to witness the presentation (after thawing).

Let's also suppose that with the help of a statistician you do the math and everything checks out. When the scale of the operation is taken into account, you calculate that instructions comparable to what you saw should pop up every 7.2 years, on average. It took a bit longer than expected, but it was worth the wait. Your skepticism has been answered. The chef was right. And interestingly, while the vast scale of repetition is what caused the instructions to be produced, what caused them to be *noticed* was selection, of a kind. So selection did have a role—more modest than invention, certainly, but not insignificant.

Of course, considering the amount of supposing we've done here, closer inspection may show this whole scenario to be implausible. We'll settle this over the next two chapters. Here, the point is that accidental invention would have to leverage repetition to beat the unfavorable odds of extraordinary coincidence. With respect to the invention of living things, then, a commitment to materialism is a commitment to accidental explanation, and a commitment to accidental explanation is a commitment to coincidence, and a commitment to coincidence is a commitment to the power of repetition. These things stand or fall together.

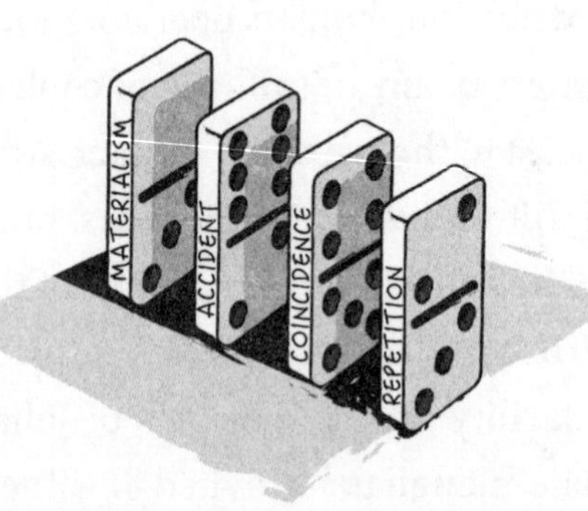

Figure 7.4 The dominoes that must stand if materialism is to stand.

IF ANYTHING POWERS ACCIDENTAL INVENTION, IT IS REPETITION

Only by improbable coincidence can accidental causes do the work of insight, and only by repetition can this improbability even conceivably be offset.

REAL SELECTION—GOOD, BAD, AND UGLY

To round out this chapter, I want to describe three case studies from the laboratory that bear out the conclusion we've drawn about selection—that it's an aimless wanderer, incapable of inventing. However strong the desire is to portray selection in glowing terms, the reality confronting scientists who work with it in the laboratory is much more humble. No one has a truer sense of what selection can and can't do than those who've attempted to harness its power, to make it perform before our very eyes. When I say that these people—myself included—have, over the decades, come to a much more modest view of natural selection, I'm saying something worth listening to.[4]

The modest view isn't entirely negative. Selection does one thing reasonably well, in fact. Having failed as an inventor, it has managed to prove itself as a *fiddler*, referring to the kind of fiddling we do in a cluttered toolshed or garage. Just as a stalled motor can sometimes be made to run with a slap on the side, or a barely working piece of equipment can be made to work better with a drop of oil here or the turn of a wrench there, so it is with biological systems. Small adjustments can sometimes mean the difference between working poorly and working well, and selection seems to have a knack for finding adjustments of that kind.

I once constructed a mutant enzyme that proved to be a fiddler's dream. Starting with a natural gene that protects bacteria from penicillin by encoding a penicillin-inactivating enzyme called *beta-lactamase,* I mutated this gene to the point where its encoded enzyme barely worked. The weakly functional enzyme enabled the bacteria producing it to survive very low doses of penicillin, with anything higher being lethal. Like a rusting motor in a junkyard, this beat-up enzyme turned out to be just the sort of thing a fiddler can fix. My colleagues and I turned selection loose on it in the laboratory by making lots of mutational variants of the encoding gene and letting selection choose which ones worked best. After six rounds of mutation and selection, using increasingly stringent selection with each round, we found ourselves with an amazingly well-repaired enzyme.[5] In fact, the five-hundred-fold improvement accomplished by the natural fiddler even surpassed the performance of the highly proficient natural enzyme I had beaten up!

Under these favorable circumstances selection is indeed able to home in on fitness to arrive at a well-tuned function. At the start of our laboratory experiment, the bacteria were in a situation analogous to a noise-seeking robot within earshot of a football stadium and with an unobstructed path to the stadium entrance. But as effective as homing was in this experiment, it did nothing that resembles *invention*. In order to make its improvements, selection had to be given a gene that encodes a working beta-lactamase enzyme, which is no small thing.

Remember that protein molecules that form enzymes must be folded into just the right shape to perform highly specific chemical reactions. The precise shape of each protein and, typically, the precise assembly of these into a multi-protein complex are what enable enzymes to perform their tasks with remarkable efficiency and precision. Selection did a fine job of making the necessary adjustments to return my messed-up beta-lactamase to good working order, but drops of oil and turns of a wrench are a far cry from the genius we associate with invention.

The best way to prove this is to challenge selection to come up with a marvel of its own. We did this too, by turning it loose on another protein that gives bacteria slight protection against penicillin. Like the weak enzyme just described, this variant was derived from the natural beta-lactamase enzyme. In this case, however, the structural disruption was so extreme that the protein didn't even qualify as an enzyme. Its encoding gene had suffered the deletion of 108 DNA bases, the loss of which prevented formation of the cleft where the chemical inactivation of penicillin normally occurs (Figure 7.5).

Nevertheless some of the simplest chemical reactions can occur even without enzymes, and inactivation of penicillin is one of them. Penicillin is a fragile molecule that breaks down in a matter of days in pure water or hours in acidic water, so nothing like the sophistication of an enzyme is really needed to inactivate it—unless you're in a hurry. Bacteria *are* in a hurry, though. Since they can reproduce within half an hour of their "birth," they can't afford to wait around for penicillin to break down on its own. Beta-lactamases reduce that wait from days or hours to minutes or seconds.

I discovered the deletion mutant after exposing bacterial cells carrying a variety of severely mutated test genes to just enough penicillin to keep them from growing.[6] Under those conditions, all that was needed for one of the cells to grow was a small enhancement of penicillin's natural tendency to break down, perhaps by something as simple as a floppy protein chain with several acidic amino acids. One mutant gene met this

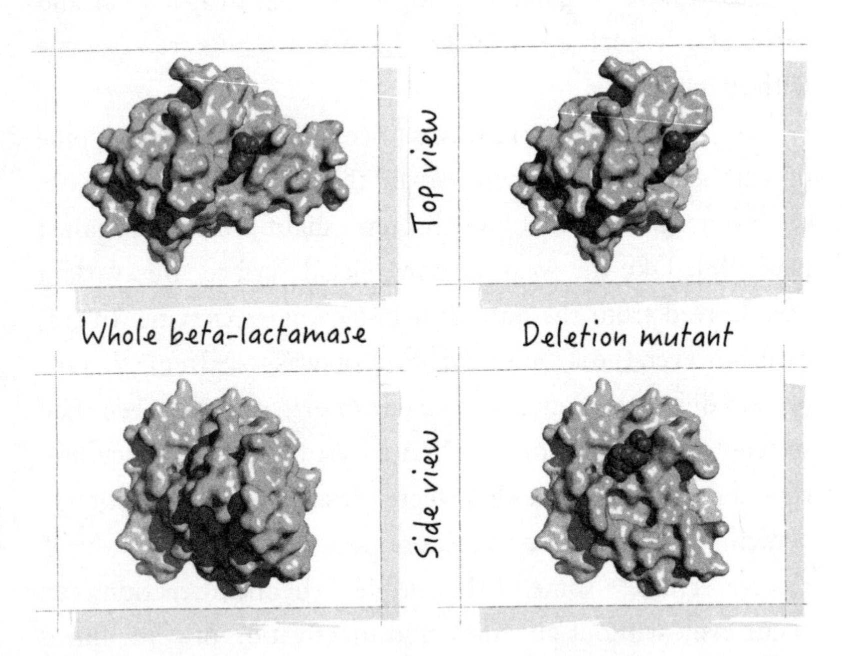

Figure 7.5 Surface renderings of the natural beta-lactamase enzyme (*left*) and the most that could remain of the natural structure in the deletion mutant (*right*). After binding a molecule of penicillin (dark gray) in the active site cleft as shown, the enzyme (*left*) rapidly inactivates it and then expels the harmless product. Once the cleft is cleared, the enzyme is ready to bind the next penicillin molecule. The image on the right is hypothetical in that it depicts what remains of the deletion mutant as if Legos had been removed from a Lego structure. Proteins are very unlike Legos, though. They tend to form their structures in an all-or-nothing way, which means large deletions like this may easily prevent the folding of what remains. Protein chains that don't fold at all remain floppy, like cooked spaghetti in boiling water. We don't know for sure whether the deletion mutant is floppy, but we know its low-level function doesn't employ the mechanism of the true enzyme on the left because it is indifferent to the removal of amino acids that are crucial to that mechanism.[7]

challenge, and while I can only guess how its encoded protein enhances the breakdown of penicillin, I was able to show conclusively that it doesn't operate the way a beta-lactamase does (see Figure 7.5 legend).

Having seen that the deletion mutant gives bacteria slight protection from penicillin, we wanted to see whether selection could leverage that effect to invent an enzyme with the structural and functional sophistication of a natural beta-lactamase. Despite our best efforts—supplying the great fiddler with all the opportunities we gave it previously—this time it failed, leaving us with an "evolved" protein that performed no better than the feeble one we had started with.[8]

Comparing this to the previous result gives a clear picture of selection's inability to invent. Homing was in operation in both experiments, but the outcome was successful only when the signal being homed in on was coming from the right kind of source. In the second experiment there was no sophisticated mechanism underlying the breakdown of penicillin, and this turns out to be much more important to the evolutionary outcome than the initial signal levels, which were similar in the two cases (Figure 7.6). Just as noise was noise to our robot, so fitness is fitness to selection, and this makes homing profoundly ineffective if the signal from the right source isn't detectable from the outset. Unless a working enzyme is supplied, the fiddling that selection does so well is useless.

There are more embarrassing examples. Selection can home in on the wrong source even when the signal from the right source *is* detectable from the outset, and worse still, it can burn all bridges to the right source in its slavish pursuit of the wrong source. This "ugly" scenario is equivalent to the robot being lured to its demise by the sound of an approaching bus, just outside a football stadium. A collaborative project between scientists at Biologic Institute and the University of Wisconsin–Superior demonstrated this by examining the evolutionary fate of bacteria carrying a faulty version of a gene that encodes one of several enzymes needed to make *tryptophan,* one of the twenty amino acids used to make proteins.[9] The faulty gene carried single DNA base mutations at two locations, each of which resulted in a wrong amino acid being incorporated into the enzyme. Both of those errors had severe functional consequences. One of them was disruptive enough to eliminate function on its own, and the other caused substantial but not total impairment. As a result, bacteria carrying this broken gene were incapable of growing unless they were given enough tryptophan to survive.

Now, you would think selection ought to have been able to repair this faulty gene as long as the bacteria were given enough

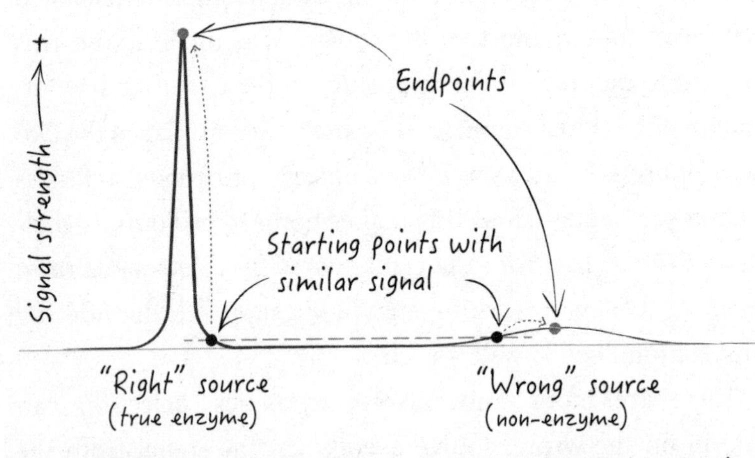

Figure 7.6 How the *source* of fitness, not the degree of fitness (the "signal"), determines the evolutionary outcome. The heavy line shows (with artistic license) the strength of the fitness signal at various distances from the two different sources. In both cases homing caused movement to the top of the local peak, indicated by dotted gray arrows. The end result was therefore determined by which source was being homed in on.

tryptophan to grow and reproduce slowly. After all, stepping stones to full restoration seem to have been carefully laid out. An initial mutation could have corrected the debilitating error, bringing the benefit of partially restored tryptophan production. That advantage should have then led to an abundance of cells with the partially repaired gene, which would have set the stage for a second mutation to correct the remaining error. Once this happened, the benefit of normal tryptophan production would have enabled the restored cells to flourish.

I refer to these contrived circumstances as stepping stones because nature would be incapable of catering to evolution the way these scientists did—supplying a nearly correct starting gene and giving the cells tryptophan handouts until they no longer needed them. So evolutionary success in this artificial scenario would do little to strengthen the case for the creative power of evolution in the wild. Failure, though, would be an example of evolution going wrong even under unrealistically favorable circumstances.

It did go wrong. Not only did selection fail to restore the faulty gene, but it also led to irreversible *inactivation* of that gene![10] Moreover, it did this by doing exactly what selection does well: homing in on the most accessible source of higher fitness. Because it costs a cell something in materials and energy to construct protein molecules by decoding the sequence instructions on a gene, faulty genes are a burden to cells that carry them. Silencing these genes so they can't be processed *at all* removes this metabolic burden. Restoring the gene would have brought a much greater advantage in this case, but selection is incapable of turning down immediate advantages for the sake of something we would consider worth waiting for. The advantage of partially

Figure 7.7 How selection went badly wrong even in a case where everything seemed poised for success. Each stepping stone represents a different version of the experimental gene, the dark spot marking the initial version with the two mutations. Stone heights represent the fitness of bacteria carrying the respective genes. The three stepping stones shown on the left seem to pave the way for selection to carry the bacterial population upward to full restoration, but experimental populations never took this path. Because the gene can be silenced by any of a great many mutations—far more than the five shown—this was the preferred outcome. Silencing mutations are evolutionary dead-ends because they typically lead to loss of the gene.

restoring the gene would have been just as immediate, but it would have been harder because it would have required one *particular* mutation—undoing the debilitating base change—whereas any of a great number of mutations were able to silence the gene. In the end, then, even hand-placed stepping stones couldn't lead selection up the right path (Figure 7.7).

The Take-Home from Homing

Having given natural selection due consideration, we conclude that it lacks the power to invent. This isn't to say that selection is completely useless—only that it's useless as an inventor. In the end, we're left with two candidates for the role of life's inventor, one fitting our design intuition and the other challenging it. If insight turns out to be the only plausible cause of invention, our intuition will have been confirmed. Alternatively, if repetition—presumably on a scale *far* beyond the familiar—turns out to have the power to invent, our intuition will have been overturned.

In order to decide between these alternatives, we need to consider the limits of repetition. Our noise-seeking robot was able to find a stadium only by wandering around aimlessly long enough to come within earshot, and we expect something similar for evolution. That is, to stumble upon an invention, a species would have to wander aimlessly long enough for this to be a likely outcome. But is this possible? To find out, we'll think about the limitations of blind searches in the next chapter.

CHAPTER 8

LOST IN SPACE

In chapter 6 we saw that our design intuition explains why we perceive busy wholes to be products of intent and how living things epitomize that category. In chapter 7 we asked whether anything has the power to defy the odds—and our intuition—by inventing life without intent. Despite all the fuss surrounding it, natural selection lacks this power. Noting the parallel between our noise-seeking robot's aimless wandering and the similarly aimless genetic wandering of a species, we realized that repetition is the only factor that can conceivably offset the improbability of stumbling upon biological inventions by accident. Having associated this improbability with coincidence, we still need to connect it to the earlier subjects of busy wholes and the universal design intuition. We'll do this in chapters 9 and 10.

In preparation for that, we turn now to the question of whether some things might be so hard to find by aimless wandering that we should consider their accidental discovery to be *impossible*. If this turns out to be true, we'll want to know whether biological inventions are among these unfindable things. If they are, we'll know Darwin was wrong.

Hunting for Eggs

The familiar way to find something is to search for it, which, in our experience, is always a goal-directed effort. Here we'll use the word *search* differently. For our purposes it will help to call *any* process that could potentially find something a search, whether or not there was a goal. In this broad sense of the word, our noise-seeking robot in the previous chapter searched for a football stadium, and evolving species search for helpful biological inventions.

We'll call searches like these *egg-hunt* searches because they have several important characteristics in common with Easter egg hunts. The first of these is that there definitely is something special to be found, whether or not the searcher is aware of it. The existence of recognizable "treasure" of some kind makes a successful outcome both possible and unambiguous. Not all searches are like that. The person combing a beach with a metal detector or sorting through a coffee can full of old coins *hopes* there's something valuable to be found, but there is no guarantee that such a thing exists.

The second characteristic of egg-hunt searches is that they occur within a well-defined space. That is, they all start with treasure somewhere *out there,* where *there* refers to a definite, bounded region. The smaller this region is, the easier the search will be, but there are no limits; the region *could* be so large that the search is effectively impossible. If a watch left on a train in

London fails to show up at the lost-and-found office by the next day, we know it's within a day's journey from London, but the portion of the globe that meets this condition is far too big to search—the watch is gone.

The final characteristic of egg-hunt searches is that they always proceed without assistance. The only way to get the treasure is to keep looking or wandering within the defined search space until it is found. There are no hints or guiding signals or anything else that systematically aids success. To our noise-seeking robot, for example, the only noise that served as a guiding signal was the sound of the roaring crowd at a football stadium. Countless other noises might have been homed in on, but none of them would have led the robot to a stadium in any systematic way. Consequently, the robot's wandering when it's out of earshot of any stadium qualifies as unguided searching. This stands in contrast to the robot homing in on the noise from a nearby stadium, or to a child finding the hidden egg with the help of a parent saying "warmer" or "colder."

Unassisted searches are often called *blind* searches. We will use this term, keeping in mind that it refers to absence of foresight or insight rather than to absence of sight. The searcher in an egg-hunt search moves through the search space, deliberately or not, and has the ability to make use of the treasure lying in this space if and when it is found, but is otherwise absolutely clueless.

SEARCHING NONPHYSICAL WORLDS

All the search examples just mentioned share a property so typical of familiar searches that we tend to overlook it: they're based upon physical location. Our robot moved from one physical location to the next, succeeding only by arriving at the physical location of a football stadium. The beachcomber checks one physical location after another, hoping one of these places will be where the desired treasure is hiding. Coins in a coffee can may be moved around freely to facilitate their examination, but the objective is still to physically locate coins that make the effort worthwhile. Even web searches come down to physical location by connecting the searcher to a physical server with the desired content.

What would a search that *isn't* grounded in physical location look like? The answer is that it would take place in the realm of *ideas*. For example, consider the game of twenty questions, where one player thinks of an object that's to be guessed by the other players. The guessers take turns asking up to twenty questions about the object, with the only permissible answers being "yes" and "no." Notice that while this game revolves around a chosen physical object, the search is not for the object itself but for the *thought* of it, expressed by naming it. Indeed, the same guessing game can be played with nonphysical categories, such as occupations or family names or songs.

The space to be searched in these games is not a physical space but rather the conceptual space of possible answers—all the answers that could, as far as the guessers know from the outset, be correct. If it makes more sense, think of the search space as an abstract set—a group of conceptual possibilities, not a group of physical things or a physical space where such things may be located (a warehouse, for example). As always, blind searches consist of checking one possibility after another—deliberately or not—while staying within the search space.

We'll soon see that egg-hunt searches in nonphysical search spaces—the kind of searches relevant to Darwin's evolutionary mechanism—are where the simple meets the surreal. Searching remains utterly simple, but *finding* becomes incomprehensively difficult as the spaces themselves become uncannily large.

This raises an important question about the meaning of *impossibility*—one we'll need to consider in order to decide whether evolutionary invention is impossible. On the one hand, because the thing being searched for—the search *target*—definitely exists within the search space, it's *theoretically* possible for a blind search to find this target. But on the other hand, since Darwin's account of invention would have to work in real life, we should reject this account if we find it to be *practically* impossible.

To give us a tangible feel for magnitudes of unlikelihood, as we explore the distinction between what is and isn't possible, let's look at a specific search that will become a helpful point of reference.

THE CUNA SEARCH

We have a better feel for physical search spaces than nonphysical ones, so our reference will be an egg-hunt search in a physical space. Because we're ultimately trying to grasp the more extreme improbabilities of nonphysical search spaces—evolutionary ones in particular—we should push our physical picture to the very limits of familiarity. The largest physical space we routinely navigate is the earth's surface, so we'll use this as our reference search space. Our search target will be a feature on the earth's surface that's just large enough to be seen when we stand above it.

I have in mind a certain indentation, about the size of a pinhead, in the middle of a particular bronze plaque that's fixed to the ground. The significance of this indentation is that it lies precisely in the "crosshairs" formed by the borders of Colorado, Utah, New Mexico, and Arizona. I'll refer to this indentation as the *cuna* target, the term being composed from the first letters of the four state names (Figure 8.1). So our reference search—the *cuna search*—is a blind search of the entire surface of the earth for this cuna target, which is a standard egg-hunt search that differs from common ones only in difficulty. The cuna target covers a mere one part in *a hundred billion billion* equal-sized parts of the earth's 510 million square kilometers of surface, making this about the hardest physical search we can mentally picture.

Having a sense for how ridiculously hard the cuna search is will be helpful when we look at much harder searches (evolutionary ones in particular). The best way to refine this sense is to perform the cuna search virtually, which you can do at the GeoMidpoint website (www.geomidpoint.com/random). This site enables you to drop up to 2,000 pins to random points all over the globe, after which you can view the pinned locations on Google Maps.[1] By zooming in on the cuna crosshairs, you'll see how close the closest pin came to hitting the cuna target. You can't zoom in close enough to see something as small as this target, but that won't matter—the closest pin will be miles away.

PINNING DOWN THE COVERAGE PRINCIPLE

To give us the visual satisfaction of hitting a target, let's do a blind search of the globe for something bigger. How about Australia? With this huge target, we expect a good number of pins in every batch of 2,000 random drops to be hits. More precisely, we expect the fraction of pins landing in Australia to be nearly equal to the fraction of the earth's surface covered by Australia, and this approximation should become increasingly accurate as more pins are dropped. So since Australia covers 1.5 percent of the earth, we should expect about 30 of every 2,000 pins to land there, 30 being 1.5 percent of 2,000. You can do the experiment to check this for yourself. When I did it, 29 pins landed in Australia, which is in line with our expectation.

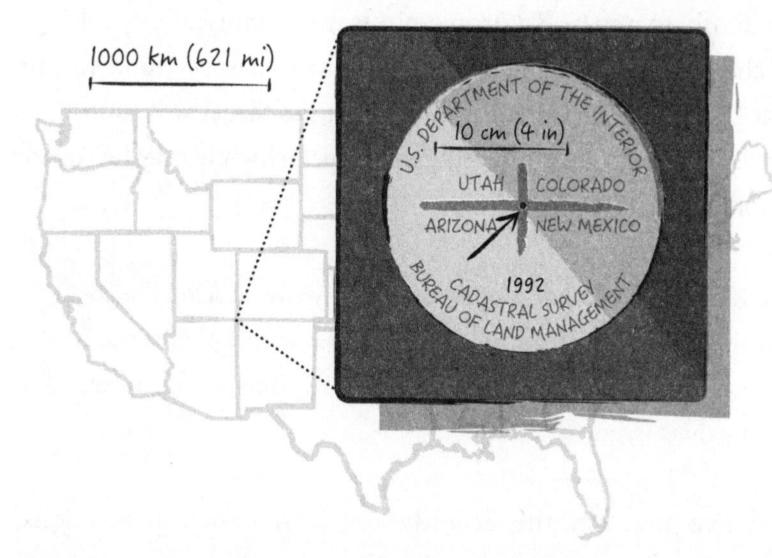

Figure 8.1 The circular bronze plaque at the center of the Four Corners Monument that marks the point where Colorado, Utah, New Mexico, and Arizona meet. The arrow points to what we're calling the cuna target, the tiny (approximately 2.5 millimeter diameter) indentation at the intersection of the "crosshairs."

The intuition informing our expectation is an obvious principle of coverage: the ease of hitting a target by accident scales with the target's size. We applied the same intuition in chapter 7 to calculate the likelihood of our noise-seeking robot landing within earshot of a football stadium, and when we found this likelihood to be very small, we appealed to repetition (repeated homing movements) as the only way to overcome the unfavorable odds. The same thing is being done here with repeated pin drops instead of repeated movements. Australia is hard to hit in one random attempt, but with a hundred attempts it becomes much easier, and with a thousand it's almost a sure thing.

We're now ready to recast our intuitive understanding as the *coverage principle,* which works just as well for a species wandering through genetic space as it does for pins dropped randomly to a map. To drop a pin is, metaphorically speaking, to check one of the possibilities in a search space, and to say that the pin hit the target is to say that the target was found. In terms of this pin metaphor, the coverage principle may be stated as follows:

> If enough pins are dropped randomly over a search space, the fraction of them hitting any target within that space is expected to equal the fraction of the search space covered by this target.

As we've just seen, this accords well with actual observations when the target is large enough to be hit easily.

In fact, this principle is so intuitively compelling that we took it to be true in chapter 7 without putting it to a test. Indeed, it *must* be true. So instead of casting it as an empirical claim, we should restate the principle as a probabilistic truth. Because probabilities are themselves fractions (fractions of complete certainty), we can replace the first occurrence of "fraction" with "probability" and remove the suggestion that many pins must be dropped:

> If a pin is dropped randomly over a search space, the probability of it hitting any target within that space is equal to the fraction of the search space covered by this target.

To apply this to our reference search, we recall that the cuna target covers one part in a hundred billion billion equal-sized parts of the earth's surface, which can be written either as a numerator over a denominator or as a decimal fraction:

$$\frac{\text{cuna target area}}{\text{earth surface area}} = \frac{1}{100{,}000{,}000{,}000{,}000{,}000{,}000} = 0.00000000000000000001$$

The coverage principle states that this fraction also represents the probability of a randomly dropped pin hitting the cuna target. We don't need a demonstration of this claim because we've deduced it from a principle we know is true.

One final adjustment will make the coverage principle more versatile. It isn't really necessary for the pins to be dropped *randomly* in order for the principle to hold. All that matters is that hitting the target isn't systematically favored in any way, which is one of the characteristics of an egg-hunt search. The pins must be dropped *blindly,* which may be quite unlike random dropping (for example, dropping in an orderly grid pattern) though it is no more conducive to success.[2] Our final statement of the coverage principle replaces "randomly" with "blindly" in order to reflect this:

> THE COVERAGE PRINCIPLE
>
> *If a pin is dropped blindly over a search space, the probability of it hitting any target within that space is equal to the fraction of the search space covered by this target.*

If I've taken you outside your comfort zone on this part of the journey, please hang in there! The terrain gets easier just ahead, and you'll see that the steep climb on this part of our hike was worth the effort. As abstract as all this talk of searches may seem, it will prove invaluable when we examine whether evolutionary searches are possible.

SPACES OF SURREAL SIZE

The coverage principle is every bit as valid for searches of nonphysical spaces as it is for searches of physical spaces. Equipped with this principle and the cuna search, we're ready to think about whether success is impossible for certain nonphysical egg-hunt searches.

Let's *try* to construct an impossible search to see whether we can. Using our intuition that huge spaces make searching harder, let's think up a nonphysical egg-hunt search in a space that's *unimaginably* large. How about the space of possible digital images? That should be plenty big. We'll have to use the word *image* loosely here, because the random pixels that fill most of this space aren't what we'd normally call *images*. With that in mind, let's take our exact search space to be this:

ALL POSSIBLE IMAGES 300 PIXELS BY 400 PIXELS IN SIZE

Leaving the search target unspecified for the moment, think about the all-inclusive vastness of this search space. For one thing, every photo that ever has been taken or ever will be taken or, indeed, ever *can* be taken has a suitably sized version in this space. Moreover, in addition to all those pictures, the space contains everything else that we'd recognize as a graphical representation of *any* kind—from circuit diagrams to wallpaper patterns to scribbled grocery lists.

The *guaranteed* existence of all this content might give the impression that we've just stumbled upon a digital treasure trove. After all, we have a container of sorts (the image space) that holds too many things of value for us to begin to number them, most of which have never been seen before. Like an archive stolen from the distant future, this container holds portraits of all the world's great leaders—past, present, and future—along with snapshots of the most newsworthy events of all time and diagrams of the best inventions of all time—innumerable priceless surprises waiting to be stumbled upon by the first explorers to search this rich space. *What a bonanza!*

Before we get too excited, though, we should remind ourselves that this is no ordinary container. Our image space is nothing more than a concept for organizing certain other concepts, namely the many possible images. And while it's true that some of these possibilities have been actualized in our physical world (Figure 8.2 being one example), it's easy to show that the overwhelming majority *can't* be. Our image space is therefore intrinsically nonphysical.

The elementary math that shows this is nothing more than multiplication of the component possibilities. Each pixel is given its hue by assigning levels (or "intensities") of the three base colors for a digital display: red, green, and blue. These levels are whole numbers, typically ranging from 0 (meaning no addition of that color) to 255 (meaning full addition of that color), for a total of 256 possible levels. The number of possible color specifications for a single pixel is therefore calculated as the product of the level possibilities of all three base colors, which comes to more than 16 million colors ($256 \times 256 \times 256 = 16{,}777{,}216$).

Figure 8.2 A digital representation of the earliest presidential portrait of Abraham Lincoln, shown at a pixel resolution of 300 (width) by 400 (height).

Because an image is nothing more than an arrangement of colored pixels, we can calculate the exact number of images in our space by multiplying those color possibilities across each of the 120,000 pixels (300 × 400 = 120,000). Taking just the first two pixels, there are 16,777,216 × 16,777,216 color combinations, which amounts to *hundreds of trillions*—already a huge number with 119,998 pixels still left to factor in! Each of these remaining pixels multiplies the possibilities by another factor of 16,777,216, producing a final number so large it has to be seen to be believed. My computing software does the full calculation in a fraction of a second, resulting in a *book-size* number—one that would take *198 pages* to print!

For comparison, a single 80-character line of text would suffice to write out the number of atoms in the universe, with the total number of physical events over the universe's history requiring only half a line more.[3] So as large and old as our universe is, it envelops *nowhere near* enough matter and has spanned *nowhere near* enough time for each of the possibilities in this search space to have been given a physical representation. The search space can never be actualized in that way, and yet it has true properties that can be verified by analysis, including the strange combination of incomprehensibility and exact calculability with respect to its size. It is at once real and *sur*real.

FANTASTICALLY BIG NUMBERS

The distinction between numbers that are so big they can't be represented physically (because there aren't enough physical things to match the number) and numbers that can be represented physically is important enough for where we're heading that I want us to have an easy way to spot the difference.

In everyday life, we think of numbers as being *big* at the point where counting to them by ones becomes inconvenient. A chaperone on a school field trip easily counts a few dozen children by keeping a mental tally, but to count hundreds would require a more elaborate process. So the dividing line between comfortable numbers and uncomfortable numbers—in this practical, everyday sense—lies somewhere in the vicinity of one hundred.

Interestingly, this common understanding of numerical "bigness" turns out to be handy when we try to get our heads around numbers that defy physical representation, which we'll call *fantastically* big numbers. As a rough rule of thumb, when the number of digits needed to write a number out is *itself* a big number, the number represented by those digits is *fantastically* big. That is, numbers exceeding about a hundred digits in length also exceed physical representation, or nearly so. For example, the number of pixel colors—16,777,216—is *big* but not fantastically big, whereas the number of images in our search space is *fantastically big*.

In fact, a much smaller search space would still be fantastically big. The space of tiny 3-by-5 pixel "images," for example, includes this whopping number of possibilities: 2,348,542,582,773,833,227,889,480,596,789,337,027,375,682,548,908,319,870,707,290,971,532,209,025,114,608,443,463,698,998,384,768,703,031,934,976.[4] Think *fantastically big* whenever you see numbers of similar or greater length, and know that these numbers are beyond physical representation.

SEARCHER VS. SPACE

We're now ready to fully describe our impossible search. We can think of any search as a contest between the searcher and the search space, with larger targets making things relatively easier for the searcher. I claim that our image search space is so fantastically big here that we can choose a fantastically big target and *still* show that the space wins the contest. If this is true, it's an important lesson to carry with us into the next chapter, where we'll consider whether egg-hunt searches can conceivably invent things.

Figure 8.3 shows one of many ways to make a fantastically large target for our impossible search. Our trick is to use pixels to represent dot-matrix characters (letters, numerals, and symbols). Fifty rows of characters with fifty characters per row perfectly fills our 300-by-400 pixel image size, turning the image into what looks like a small phone screen packed with characters. Let's call any image that's filled with characters in this way a *dot-matrix screenshot,* just to give it a name. Our *search target,* then, will consist of:

ALL IMAGES THAT DEPICT *ANY* DOT-MATRIX SCREENSHOT

We know the number of these target images will be at least as large as the number of possible character combinations, which is staggeringly large. We can make it even larger, though, by not requiring each character to be made of black pixels against a white background. By experimenting, I found characters to be readable if their color levels (red, green, and blue) are in the lower third of the range (0 to 85) while the levels for background pixels are in the upper third (170 to 255). Now, instead of using only two pixel colors (black or white) to build the dot-matrix characters, we can use over a million.

So our contest looks like this: the searcher will check as many images from the search space (all possible 300-by-400 pixel images) as possible to see if any happen to "hit" our search target by being a dot-matrix screenshot. Exactly *how* this is done doesn't concern us, as long as the process is truly blind, meaning the choice of possibilities to check doesn't in any way benefit from insights that would favor correct guesses. If you want to picture the process, think of a web service that allows the searcher to upload an unlimited number of 300-by-400 pixel images with an email notifying the searcher immediately if one of these uploaded images happens to be a hit.

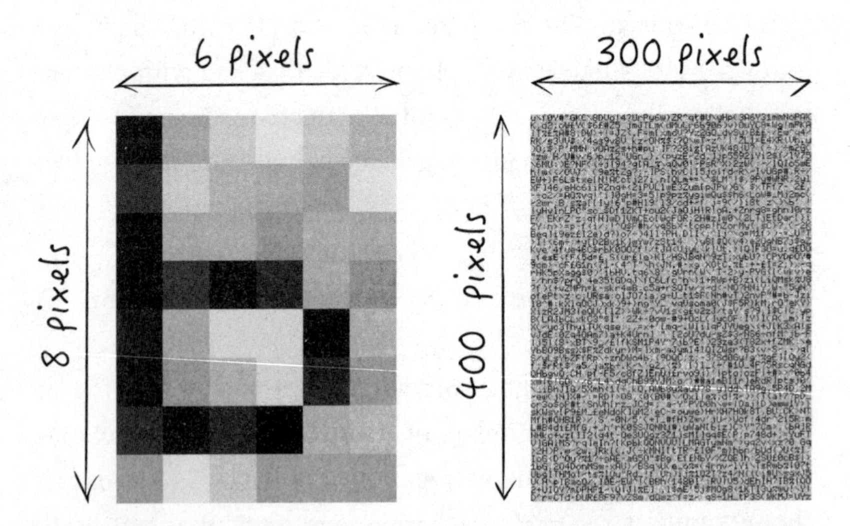

Figure 8.3 Light and dark colors (shown in grayscale here) are used to turn a 300-by-400 pixel image into what we're calling a *dot-matrix screenshot* (50 rows of 50 characters). Each character occupies a 6-by-8 pixel rectangle (*left*). I used a set of 93 characters (upper and lower case letters, plus numerals and symbols) in random order to construct the example of a dot-matrix screenshot on the right.

That's just one way of picturing the search, though. It could look very different. The key point is that no matter what the search looks like, every guess consumes physical resources. Physical things have to be manipulated for a guess to be constructed and tested, and this requires at least a little bit of physical material to be devoted to representing each guess for a little bit of time. No matter how efficient or extensive the things that make and check guesses are, these little bits become huge as the number of guesses becomes huge. And as abundant as physical materials and time are, they aren't *infinitely* abundant. So there comes a point where the search is so demanding that it simply can't be completed, no matter how the searcher tries to approach it. In the final analysis, then, the contest boils down to a comparison of *a,* the number of images that can be physically actualized (so they can be checked), with *b,* the number of images that would *have* to be actualized in order for a dot-matrix screenshot to be among them by chance.

We know *a* can't be fantastically big because we've defined fantastic bigness to mean *too big to be actualized.* This means the searcher is in trouble if *b* is fantastically big. We can estimate *b* by using the coverage principle along with the commonsense rule that it takes about a gazillion tries for a one-in-a-gazillion outcome to happen by chance ("gazillion" here stands for any specific large number). This way of estimating *b* is worth stating in terms of our pin-dropping metaphor because we'll use the same method to decide whether evolutionary searches are feasible. Using "reciprocal" in the mathematical sense, where the reciprocal of *m*/*n* is *n*/*m*, we have:

> THE PRINCIPLE OF RECIPROCAL SCALE
>
> *The number of pins that must be dropped over a search space before a particular target is expected to be hit blindly can be estimated as the reciprocal of the probability of success on the first drop, or—equivalently—the reciprocal of the fraction of the search space covered by this target.*

Translating from the pin metaphor, this means the number of images the searcher will have to check before we'd expect a dot-matrix screenshot to be among them is equal to the size of the space divided by the size of the target:

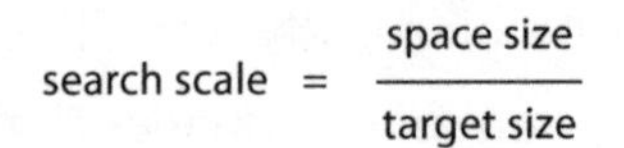

$$\text{search scale} = \frac{\text{space size}}{\text{target size}}$$

This puts the frighteningly big 198-page number in the numerator, which means the searcher can only hope we've made the target large enough for the resulting answer to not be fantastically big.

Instead of exhausting you with more multiplication, I'll just say that the number in the denominator—the number of possible dot-matrix screenshots—is also fantastically big, filling about 160 pages. As huge as that is, though, it's not nearly big enough to do the searcher any good. The trick for gauging the size of the answer is to subtract pages, as in Figure 8.4. This tells us that the answer—the number of images that would have to be actualized—is a *38-page* number, which we know is fantastically big. The fact that this is much smaller than it *could* have been (had the target been smaller) is of no consequence. The blind searcher *can't* succeed because success would require more images to be checked than any physical process *can* check. The search space wins this contest decisively.

We can use the cuna search to get a feel for just how hopeless the situation is for the searcher. The principle of reciprocal scale tells us that the expected number of blind pin drops needed to hit the cuna target is about a hundred billion billion, which is written out as a one followed by twenty zeros. For a run of *n consecutive* cuna hits (if you can imagine such a thing), the number of blind drops needed is therefore expected to be a 20*n*-digit number, where *n* is the number of hits in a row. The number of blind pin drops needed for four cuna hits in a row, for example, would be an eighty-digit number, which would fill a line of text, without commas.

We now have two ways to gauge the difficulty of very hard searches. Both make use of the huge number we get by applying

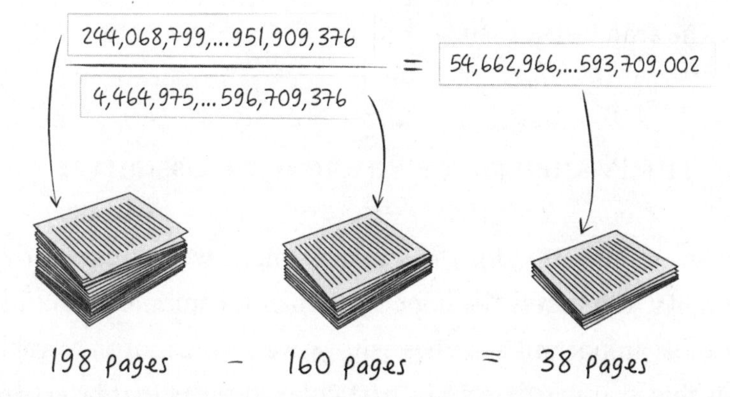

Figure 8.4 Division of extremely large whole numbers (larger divided by smaller) is represented in abbreviated form at the top, with the full length of each whole number depicted by a stack of printed paper below. For numbers that aren't of page length, you can use the same method with *lines* instead of pages. Notice that no division has to be performed to gauge the magnitude of the result in this way.

the principle of reciprocal scale and in particular the number of digits needed to write this huge number out. First, dividing the number of digits by twenty tells us how hard the search is in terms of consecutive cuna hits. So, if the principle of reciprocal scale says the number of possibilities the searcher needs to check is a forty-digit number, this means the search is as difficult as dropping pins blindly until the cuna target is hit twice in a row—*astonishingly* improbable, when you picture the cuna search. Second, we know that if the number of digits is above a hundred, then a *fantastically* big number of possibilities would need to be checked, which simply can't be done.

At thirty-eight pages, the whopping number of images that would have to be checked for the searcher to find a single dot-matrix screenshot has over 160,000 digits. Dividing this by 20, we find our blind image search to be as difficult as dropping pins blindly until the cuna target is hit *eight thousand times in a row by pure luck!*

That will never happen.

The search space wins.

The Possibility of Physical Impossibility

Anyone still rooting for the searcher might seek refuge in two thoughts. The first is the hope that when it comes to evolution, the most important searches will prove much more favorable than this example search. In particular, if the actual targets of interest cover a substantially higher proportion of their respective spaces than this one does, perhaps the principle of reciprocal scale won't ultimately be an insurmountable obstacle. This certainly has to be given due consideration before we reach any firm conclusions regarding evolutionary searches, a task we'll tackle in the next chapter.

The second potential place of refuge is the thought that the word *impossible* ought to be reserved for situations where the probability of success is exactly zero. Admittedly, this isn't true for our example. Rather, the probability of getting a dot-matrix screenshot by chance in one try would be represented as a zero followed by a decimal point, followed by a *very* long run of zeros—filling thirty-seven pages and spilling over onto the thirty-eighth page before the first non-zero numeral appears. That probability can be increased by allowing more tries, but the whole point is that tries can only be multiplied within hard physical limits. Even under the most wildly optimistic assumptions, our universe—vast and ancient as it is—can't muster enough repetitions to erase more than about a hundred of those zeros!

Keep in mind that our interest here is more practical than mathematical. Students of math should, of course, learn the conceptual distinction between infinitesimal fractions and zero. But to decide whether success is possible enough to carry any real implications is to make a *practical* distinction, not a conceptual one. Bearing that in mind, it's clear that some search challenges favor the search space over the blind searcher so overwhelmingly that they should indeed be regarded as impossible. More precisely, success should in these situations be regarded as a *physical* impossibility in order to distinguish it from a conceptual impossibility. We're free to tell stories about such long odds being beaten, but we now see very clearly why tales of that kind belong in the fiction section—where we filed the tale of oracle soup.

Whether Darwin's account of life belongs there too remains undecided for the moment. If the dominoes fall, his theory falls with them. Repetition would be the first one to topple, should it prove inadequate for explaining the remarkable coincidences needed for life to be an accident. And it may. As we have now seen, that domino is rather wobbly.

CHAPTER 9

THE ART OF MAKING SENSE

Aiming to resolve the conflict between our design intuition and the evolutionary story, we set off on a quest for understanding that has been like a hike up a mountain trail. We started, at sea level, with an idea so familiar it has the feel of something obviously true. This was the universal design intuition. Lingering at low elevation for a while, we took time to appreciate the humanness of science before starting our climb. Pressing upward, we eventually found ourselves in the rarified air of the summit, where we encountered subjects that may have seemed quite unfamiliar. The walk will be easier from now on because we have reached the highest point. We have only one or two more things to see at this elevation before we work our way back down to level ground, revisiting along the way places we've seen before.

On the approach to the summit, I led us on an exploration of the general subject of blind searches. As challenging as that topic may have been, it will prove critical for sorting out the conflict between the evolutionary story and our design intuition. In fact, the evolutionary process as described by biologists is really nothing more than a large collection of ongoing blind searches—one for every species in existence. There's nothing controversial about this. By *blind* I mean without foresight or understanding, just as evolutionist Richard Dawkins did in his highly acclaimed defense of Darwinism, *The Blind Watchmaker.*[1] And by *search* I don't mean anything inconsistent with complete blindness. The idea is not that any species *aims* to acquire new features but that all species *do* acquire new features, supposedly, through a long process of genetic meandering similar to the meandering of our noise-seeking robot in chapter 7. So it's correct to say that any remarkable biological features acquired in this way were *found,* not by deliberate effort, as the hound finds the fox or as the detective finds the murderer, but rather by ordinary course of nature, as the river finds the ocean or as the lightning bolt finds its path to the ground.

The contrasting view is that what looks to be the fruit of genius always *is* the fruit of genius. The universal design intuition declares this to be so, and everything in our daily experience affirms that declaration. Inventions are clever things, and clever things are to be had only by cleverness. Inventors do sometimes search for new ways of doing things, but they never search *blindly.* Invention is, after all, about mental lightbulbs going on so that things are seen clearly for the first time. It's nothing like the aimless groping in the dark that characterizes a blind search.

How Humans Invent

Without presuming that all invention must proceed the way human invention does, it will help to think about how we humans invent. I realize that most of us don't think of ourselves as inventors any more than we think of ourselves as scientists, but this is because we underestimate the significance of what we do with routine ease. When we get to the bottom of what invention really is, we'll see that inventing is an essential part of being human.

To help us get there, I've broken down the process of human invention into the three stages shown in Figure 9.1. The first of these, the *mental* stage, is where the initial idea for the invention is developed into a detailed plan ready to be implemented. The big top-level idea must always be broken down conceptually into smaller ideas, which may have to be further broken down before implementation can begin. The downward steps in the first stage of Figure 9.1 are suggestive of this progression from the high-level concept to the nitty-gritty details.

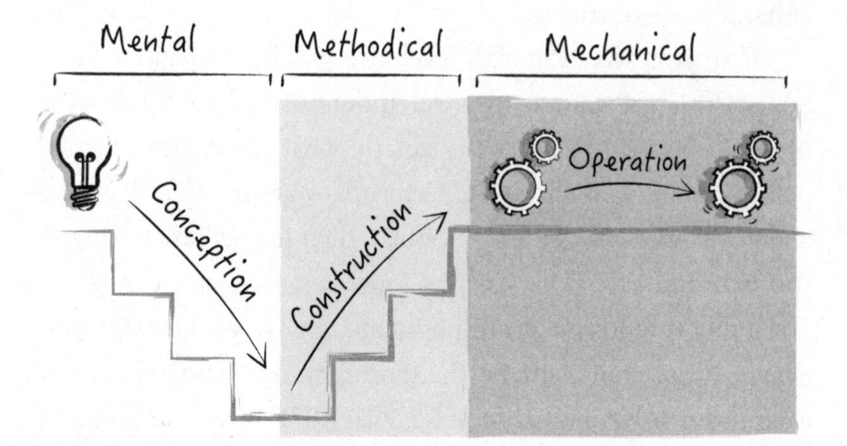

Figure 9.1 The three stages through which human invention proceeds. Shading indicates the transition from purely mental activity (unshaded) to purely physical activity (dark shading). The mental and methodical stages aren't as cleanly divided as this depiction suggests, and yet there is a real progression from the purely mental activity of conceiving to the more physically constrained activities of building and testing.

The second stage is where the resulting conceptual plan is used to construct a physical thing. The mental work at this stage is more practically oriented than it was in the first stage. The fully conceived plan is imposed on actual physical materials, which requires both a conceptual understanding of the plan and the ability to resolve all the matters of detail that arise when complex plans are implemented for the first time. I describe this as the *methodical* stage, to suggest both that it requires conscious, purposeful action and that this action must take careful account of the properties of the physical objects and materials being manipulated. Notice that as many steps are ascended in this second stage as were descended in the first. The point of the first stage was to form ideas in a top-down manner, going from the high-level idea to the low-level details needed to implement it; the point of the second stage is to form a physical device in a bottom-up manner, going from the raw materials and supplies at hand to a working prototype.

If all goes well, the high-level idea will be evident to everyone at the third stage as they watch the finished device do what it was designed to do. This is exactly what we inferred for the pool robot back in chapter 6. Everyone watching that device in action realizes it's cleaning a pool, which immediately triggers the recognition that it was consciously *intended* to clean pools. So having directly observed physical activity, we infer that past *conscious* activity produced the special kind of physical activity we're witnessing, namely the busyness of a busy whole. Upon witnessing the invention in operation, we infer that it was constructed according to a conceived plan.

An example will help to solidify these ideas.

HOUSTON, WE'VE HAD A PROBLEM

In April of 1970, NASA's Apollo 13 mission sent three men—Jim Lovell, Jack Swigert, and Fred Haise—into space with the objective of landing on the moon. On the third day of their journey, things went horribly wrong. An oxygen tank exploded, causing major damage to the spacecraft and forcing Mission Control in Houston to replan the remainder of the mission around a new objective: to bring the crew safely back to Earth. Despite the precarious start, Apollo 13 would become a resounding success with respect to this new objective.

Of the many critical challenges that had to be overcome in the days between the explosion and the eventual splashdown in the South Pacific, one was to prevent lethal buildup of carbon dioxide (CO_2) in the section of the spacecraft where the astronauts were living, called the *lunar module*. Scrubber cartridges were present onboard for that purpose, but the ones that were accessible to the astronauts were box-shaped, for use in the command module, whereas the lunar module was designed to use cylindrical cartridges. With the crew's lives at stake, engineers on the ground had to come up with a way to make the box-shaped cartridges work with a system that was designed to use cylindrical cartridges. Their now-famous solution to this challenge was nicknamed the "mailbox" (Figure 9.2).

Like all inventions, the Apollo 13 mailbox had its origin in thoughts. First came the motivating thought, which was the realization that CO_2 exhaled by the astronauts would become lethal if nothing were done to remove it. Next came an analysis of the situation that provided the most promising path for a solution. So before any physical things were manipulated, *ideas* were manipulated and refined with the aim of thinking through all the details needed for the big idea to succeed. This is the first stage in Figure 9.1—the mental stage.

Early in this process, the engineers on the ground started to progress into the second stage of Figure 9.1—the methodical stage of construction. I say *started* because the mental and methodical stages of invention usually overlap. It's almost always necessary to experiment with ideas by trying them out, and the ideas are almost always refined in the process. As important as experimenting is, though, the urgency of the astronauts' predicament demanded a speedy conclusion. Up in space, where it really mattered, everything was riding on the final stage of invention: the stage where the mailbox must prove its worth by working. The astronauts therefore needed to bypass all the head scratching and experimenting their colleagues on the ground were engaged in and, after one success-

Figure 9.2 The Apollo 13 "mailbox," seen toward the top of this NASA photo, was a jury-rigged invention for removing CO_2 from the lunar module. Without it, the astronauts would have perished.

ful pass through the methodical stage of construction, move straight to the final stage.

To pull that off required an invention of another kind: a clever arrangement of words called *instructions.*

LANGUAGE, THE ULTIMATE MEDIUM OF INVENTION

Interestingly, the importance of crafting words wasn't lost on the NASA engineers, as these excerpts from the Apollo 13 air-to-ground voice transcript clearly show:

03 08 22 13/ Mission Control:

Yes. We wish we could send you a kit and it would be kind of like putting a model airplane together or something. As it turns out, this contraption will look like a mailbox when you get it all put together.

. . .

03 10 52 51/ Mission Control:

. . . as you know, we've got a way to use those [cartridges]. And as soon as we get them written in some good words, why, we'll pass that along. You might be able to make one.

. . .

03 11 51 58/ Mission Control:

. . . we're getting the words together to make it easy to build one of these things, and it looks like it will probably take two guys, so, I think we probably ought to plan to do that later.

. . .

03 18 08 43/ Mission Control:

```
Okay, Jim. The way I thought it might be best to do it would be
to have you gather the equipment and let us talk you through
your procedure while you do it. Now, maybe you could give Jack
the headset and - and - get the equipment together, and we'll
talk you through the procedure. I think it'll be a little
easier to do that way than if you tried to copy it all down -
and then go do it.[2]
```

The spoken instructions that followed this recommendation fill many pages of the typed transcript, indicating that this was one of those situations where the details matter.

Considering how categorically different *words* are from the assorted objects Jim Lovell was asked to gather—scrubber cartridges, duct tape, plastic bags, and cardboard—it's remarkable that the same principles of invention apply across these categories. Whether we're inventing instructions or a mechanical device of some kind, we always start by conceiving, and the process of conception always works its way from a whole concept—the big idea—down to the low-level details that must be resolved for this idea to be implemented. However fitful the transition from conception to implementation—however many takes and retakes it may require—we end up with a physical thing whose definite hierarchical structure reveals our thought process. That is, everyone can see how we thought about the problem by examining the hierarchical structure of the invention we came up with to solve it. Indeed, this very structure is what causes the invention to work.

Having said that, I should add that the most elegant inventions perform their top-level functions so impressively that the lower-level functions usually go unnoticed. When the tennis player of chapter 6 wins her match, the talk is about the skill of her game, not about how well her lungs or her heart performed. Yet the fact that no one had occasion to think about her respiration or her circulation shows how proficiently those necessary physiological functions *were* performed. They supported excellent tennis so well as to make themselves practically invisible.

What enables inventions to perform so seamlessly is a property we'll call *functional coherence.* It is nothing more than complete alignment of low-level functions in support of the top-level function. Figure 9.3 illustrates this schematically for a hypothetical invention built from two main components, both of which can be broken down into two subcomponents, each of which can in turn be broken down into elementary constituents. Horizontal brackets group parts on a given level that form something bigger one level up, with the upward arrows indicating these compositional relationships. Notice

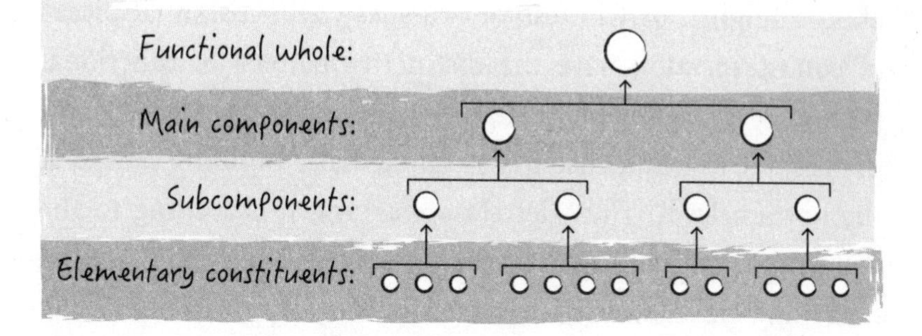

Figure 9.3 The hierarchical structure of an invention, showing the functional coherence that characterizes the relationships between parts. In this scheme the parts at intermediate levels (between the elementary constituents and the functional whole) are referred to as *components*. The number of intermediate levels and components depends both on the invention and, to a degree, on the way we choose to delineate its principal parts. The invariant fact is that the many parts must perform their small functions in a particular hierarchical way in order for the whole invention to perform its large function.

that every part functions on its own level in a way that supports the top-level function. This complete unity of function is what we mean by functional coherence.

> **functional coherence:**
> the hierarchical arrangement of parts needed for anything to produce a high-level function—each part contributing in a coordinated way to the whole

As abstract as that may sound, it has very concrete and familiar implications, all of which will be apparent when we consider how instructions work. The Apollo 13 mailbox instructions will continue to serve as our example, but instead of concerning ourselves with how CO_2 can be removed from air, we'll focus on how ideas are conveyed by language. Whether we examine written or spoken communication, and whether we choose English or Chinese or Malagasy, all substantial pieces of communication have the distinctive pattern of functional coherence represented in Figure 9.3. Alphabetic written languages, for example, use letters as the basic building blocks at the bottom level. These letters are arranged according to the conventions of spelling to form words one level up. To reach the next higher level, words are chosen for the purpose of expressing a thought and arranged according to grammatical conventions of sentence structure in order for that thought to be intelligibly conveyed.

Whether still higher levels have to be reached depends on the objective. If the point is to convey one simple thought, a sentence should suffice. If it's to carry readers through an extended thought process, many sentences will be needed, each carefully crafted to make its own point in a way that coheres with the preceding points and paves the way for subsequent points.

MAKING SENSE WITH LETTERS

Mission Control's objective of enabling the Apollo 13 astronauts to construct a device for CO_2 removal called for many instructive sentences, which in turn called for knowledge, not only of how to construct the device but also of how to put that understanding into "good words." According to our design intuition, instructions like these can only come from someone who has a mental grasp of the procedure being conveyed and of the language in which it's to be conveyed.

We're now beginning to see why this intuition has to be correct. To guide our thinking here, let's consider a hypothetical scenario. Suppose that instead of receiving spoken instructions from their colleagues in Houston, the astronauts had to work from written instructions. And suppose these written instructions had to come from a source with no understanding of the astronauts' predicament, no clue as to how it might be remedied, and no comprehension of language. Imagine the responsibility for these critical instructions resting in the prehensile hands of a monkey—perhaps a retiree from the former space-monkey days. Equipped with a typewriter but otherwise woefully ill-equipped, this monkey clearly would have had very little chance of producing anything resembling adequate instructions.

That much is obvious. More interesting is whether *any* blind search process—whether employing monkeys or supercomputers, whether operating over days or eons, whether confined to a spacecraft or distributed over all the planets in a billion galaxies—can produce enough sequences of alphabetic letters for one of them to be an effective set of instructions for building the Apollo 13 mailbox. To help us answer this, I converted the original spoken instructions into a concise written form that, at thirty-two lines, fills just over half a page. Although this isn't a lot of text, we know from chapter 8 that the corresponding space of raw possibilities—the total number of *possible* ways to fill thirty-two lines—is so large that it's physically impossible for all of them to be actualized. The question is whether the number of alternative ways to word the instructions might conceivably be large enough for some blind search process to triumph over this enormous search space. That proved not to be the case for the contest between searcher and space that we examined at the end of the previous chapter, but we followed that conclusion with the thought that perhaps the most relevant targets do cover enough of their spaces for a blind search to succeed.

Certainly, everything that can be conveyed in words can be conveyed in many ways. As for *how many* ways, though, I know of no way to do the counting. Thankfully, we can evaluate this search by a different approach. Instead of trying to count the alternative wordings for mailbox instructions, we'll simply ask how rare the functional coherence is that all written instructions require. Starting at the bottom level, where letters are strung together to make words, the first question is: how rare are the letter combinations used in writing when compared to *possible* letter combinations?

To answer this, picture two printed pages, one half-filled with intelligible writing and the other half-filled with random typing. Both pages are covered by sheets of black paper with several small rectangular holes in them, each hole large enough to expose just three consecutive letters. Would what we see through these holes reveal which page is which? If so, then coherence is discernible even in fragments of text smaller than an average word.

Figure 9.4 shows examples of what we'd see. Inspecting the four columns of letters revealed on the upper page, we see no examples that look as though they can't have come from intelligible writing. Combinations like "`ngt`" and "`rtr`" (both in the first column) might stump us for a moment, but we can easily believe these came from English words even if we struggle to call any examples to mind (for the curious, they came from *length* and *cartridge,* respectively). By contrast, the lower page reveals many obviously implausible combinations, including "`qmf`," "`xdc`," "`wvw`," "`wrm`," and "`hzj`," among others. On this lower page we also see a preponderance of letter combinations that, while not definitely implausible, are nevertheless peculiar. The "`ftv`" sequence, for example, looks as though it could perhaps exist in a compound word like *softball,* but it doesn't exist in the 93,000-word software dictionary I'm using. So it seems we *can* spot the incoherence of random typing even in tiny fragments of keystroke sequences, provided we have several fragments to inspect.

Figure 9.3 helps us understand this. In this case the elementary constituents at the bottom level are the twenty-six letters of the alphabet plus the space character for separating typed letters into words. We perceive letter combinations like "`hzj`"

of	fo	d b	g a
ge	s h	e f	uts
ti	ise	of	in
co	ros	tic	tap
ure	t a	s t	ts
par	par	th	ti
ngt	pr	an	w c
of	ap	gai	cu
nch	om	le	e c
rtr	fo	sid	e f
off	str	m f	g i
tap	e c	of	sho
se	ta	the	y c
sh	e i	he	se
fr	the	th	bag
out	ne	ho	und
g a	ins	n t	ose
pre	the	bo	t t
cut	sti	of	an
en	ght	ca	ou

qmf	tqm	xdc	bem
hmm	i a	x x	r
jh	wq	ya	ofk
wk	ly	pac	yvz
lnz	t	fsc	qvf
vuu	c u	tnc	u s
ovu	exq	uvc	c e
gnx	mfn	uzo	wrm
fb	pki	tza	gmd
lo	nnu	wvw	u
urt	e p	rxo	fsf
p	e j	juy	sqs
rde	wrm	noa	ypn
r	btq	jnq	acw
ghe	ftv	svp	vw
frh	cn	upr	dwi
hzj	zgq	ite	t p
rma	ll	bfl	nr
wng	mti	ezz	nx
xft	l k	yuq	ycn

Figure 9.4 Examples of what would be seen through the rectangular holes described in the text. Holes expose three consecutive letter positions (some of which may be occupied by spaces) at random locations on the two pages. The upper page has actual mailbox instructions. The lower page has random typing, simulated by representing the twenty-six alphabetic letters and the space character in proportion to the size of their respective keys on a typical keyboard, where the space bar is five times the size of a letter key.

to be incoherent because our familiarity with English tells us they can't form part of any English word. In terms of Figure 9.3, these combinations can't be placed under a horizontal bracket at the bottom level because they can't form a word at the next level up.

As you'd expect, the prevalence of this problem makes words a rare occurrence in random typing. Of the 248 letter groupings on the lower page shown in Figure 9.4, only eight are recognizable words. Most of these are either the one-letter words *a* or *i* (signifying *I*) or two-letter words like *he* and *uh*. The longest word on the page happens to be the three-letter word *ink*. In all, these little words make up a mere 1 percent of the page's content.

Having considered none of the finer points of writing yet, we already have what we need to decide whether accidental mailbox instructions are within the realm of physical possibility. The observation that 248 letter groupings resulted in only eight actual words means that the random letters falling between successive spaces have only about a 1-in-31 chance of being words (8/248 = 1/31). And when they do happen to be words, they tend to be very short, averaging only about two letters or three keystrokes (counting the space that ends the word). So because it takes about 1,800 keystrokes to fill half a page, random typing would have to produce about 600 consecutive words just to fill half a page with words (1,800 ÷ 3 = 600). If it did, there would be no coherence above the level of words, but the avoidance of unrecognizable letter sequences would at least satisfy that bottom-level requirement for useful instructions.

As inadequate as this requirement is, it provides an easy way to calculate a probability that can be used with the principle of reciprocal scale from chapter 8. Doing this will tell us whether a blind search of keystroke combinations can find even something as insignificant as a jumble of tiny words that fills half a page. If this overly generous target can't be found, then finding coherent instructions for building the Apollo 13 mailbox is completely out of the question.

To calculate the probability that half a page of random keystrokes would consist entirely of English words, we start with 1 and multiply by 1/31 (the probability of a letter grouping being a word) over and over, a total of 600 times. According to the principle of reciprocal scale, the number of half pages that would have to be filled with *blind* typing[3] in order for one of them to consist entirely of words is expected to be roughly equal to the reciprocal of this multiplied fraction. Equivalently, we can start with 1 and multiply by 31 (the reciprocal of 1/31) over and over, 600 times. When that calculation is done, the printed result fills just over eleven lines with numerals, making this a paragraph-size number instead of a book-size number—fantastically big nonetheless.

Living as we do in a universe that can't produce two lines' worth of physical attempts at *anything*, this eleven-line number delivers an overwhelming victory to the search space. In terms of the pin-dropping metaphor, the difficulty of the blind search finding even this meaningless jumble of short words equates to that of blindly hitting the cuna target *forty-four times in a row* (four hits per line, as noted in chapter 8).

Notice how comprehensively the blind search has been defeated. We asked whether it could produce instructions for building the Apollo 13 mailbox, and in the process of deducing that it can't, we discovered something much more profound: a blind search can't produce *any* coherent piece of extended writing at all! *Nothing* that puts half a page to good use is physically possible, whether instructions, or recipes, or to-do lists, or love letters, or poems, or anything else.

IMPOSSIBLE COINCIDENCES

The best medicine for anyone wanting to find a way around this hard fact is a clear understanding of why it really is a hard fact. The most common way to imagine going around it is by what we called *stepping stones* in chapter 7. That idea is certainly tempting in the present context. We often communicate in short phrases—*Over here!*—or even single words—*Help!* So, since blind searches can find simple targets like these, we tend to be sympathetic to the idea that successes on this modest scale, where very little functional coherence is needed, can be built upon gradually to produce successes on much larger scales—even the scale of complete instructions.

I think our sympathy has to do with the fact that we, as creative thinkers, love the idea of building on modest beginnings. We do this all the time—but not without insight. The problem is that we have no way to turn insight *off*. Insight comes so naturally to us that we supply it all the time without noticing, even when it doesn't belong. In this way we tend to help evolutionary stories along the same way we help any other story along: by filling in the gaps and adding a favorable interpretation.

Whether or not sympathy is the explanation, stepping-stone logic is very common in evolutionary discussions. For that reason I want to stress again why it doesn't work. As I said in chapter 7, blind causes are so fundamentally unlike insight that any instance of them *looking* insightful would be coincidental. Coincidences do happen, of course, but we know from experience that major ones are much more rare and therefore more surprising than minor ones. This is common science at work. A three-letter word appearing in alphabet soup is worth mentioning, a five-letter word is worth photographing, and a seven-letter word is downright suspicious.

All we're doing in this chapter is unpacking this intuition to show why our firm sense that certain things can't happen by accident is absolutely correct. What we're seeing is that the amount of functional coherence routinely produced by human insight truly *can't* be produced by accident. The reason connects to what we learned in chapters 7 and 8: accidental causes mimicking insight on this scale would be a fantastically improbable coincidence, which means a *physically impossible* coincidence.

Because the degree of coincidence is what makes accidental explanations implausible, there's no way to alleviate the problem by thinking up creative coincidental stories. However creative we make these stories, the creativity only dresses up the coincidence. None of these stories *remove* the coincidence because the very claim that accidental causes did what insight does *is* the coincidence. The problem lies with coincidence *itself*, and that's why twists in these stories, whether stepping stones or anything else, never help. Thinking back to the hypothetical team of physicists in chapter 7—this is why we didn't need to know what they meant by "correlative entrainment." As long as they meant something lacking insight, we knew they were banking on an impossible coincidence.

The implications for invention are clear. If the invention of a working *X* is a whole project requiring extensive new functional coherence, then the invention of *X* by accidents *of any kind* is physically impossible. Why? Because for accidental causes to match insight on this scale would be a fantastically improbable coincidence, and our universe simply can't deliver fantastically improbable coincidences. The fact that much simpler things can be had by accident is completely irrelevant. The only thing we need to know to reject all accounts of *X* itself being invented by accident is that these stories all attempt to excuse an impossible coincidence.

By now, none of this should sound new. Whether we speak of impossible coincidences or impossible searches, the hard fact is exactly the same: high-level functional coherence can't be found by any blind search because this would amount to an impossible coincidence. Only insight can hit a target like that, which is no coincidence.

MAKING SENSE WITH WORDS

Although we encountered this hard fact by looking at coherence at the low level of letter combinations, the situation only gets worse as we move up the hierarchy. Automatic spelling correction didn't exist in 1970, but if we imagine it did, even with a high-powered version that converts random keystrokes into the closest words, the astronauts would have been no better off. Like letters, words must be arranged coherently, which involves choosing good words and putting these words in good order. It isn't as easy to calculate the likelihood of this happening blindly as it was for forming words from letters. Still, we can easily see that vocabulary is tightly constrained by the writing objective.

For example, of about fourteen thousand seven-letter English words, my version of the written instructions for the Apollo 13 mailbox uses only *eleven*. No doubt other words could have been used, but not just any words. To get a feel for how strongly the subject of the writing constrains the vocabulary, try giving someone the words I used (*against, another, between, corners, cutting, lengths, outside, plastic, screens, secured,* and *tightly*) and asking them to guess the subject of the writing from which these words came. They'll easily deduce it has to do with some sort of construction project—one involving plastic, screens, cutting, and tight securing. The ability to make that much sense from such small pieces of a text is a mark of coherence.

If coincidental coherence is as rare at this level of vocabulary as it was at the level of spelling, you'll get a very different result when you present someone with a *random* selection of seven-letter words. One example should be enough to make the point. Here, then, are eleven seven-letter words chosen randomly from the 93,000-word dictionary that came with my computing software: *luffing, dickens, numbers, inbound, roofers, incisor, overlap, Brownie, genomes, avenged,* and *tallier.* In these words I submit there is no hint of a coherent theme.

We need go no further. Blind searching fails at all levels. As people who write, we know that the need for insight grows as we move up the hierarchy, and that only makes the coincidence of blind coherence greater and greater. Any process that can't substitute for competence in either spelling or vocabulary certainly can't substitute for competence in grammar or composition. Our design intuition has this one exactly right. We need knowledge to write useful instructions, and no accidental process can replace that knowledge.

Without belaboring the point, I want to show you how *general* this conclusion is by taking a quick look at an example that's very different from language.

MAKING SENSE WITH PIXELS

For this we return briefly to the subject of digital images, this time focusing on photographs. The pattern of hierarchical functional coherence is present here as well. Just above the bottom level of pixels, digital photographs show coherence analogous to a painter's brushstrokes, where colors are extended and blended. Above that is a level where boundaries and shapes are defined. Still higher is the level where features and objects are recognized, and above that comes the level where the principal subject takes full form, along with the setting in which it was photographed. Noteworthy photographs exhibit an even higher level, where the *way* in which the subject was photographed evokes an impression that goes well beyond mere recognition.

The universal design intuition assures us that none of this happens by accident, and again we can use the principle of reciprocal scale to confirm this. Using a collection of low-resolution photos (400 pixels by 300 pixels), I wrote a program that repeatedly picks one at random and copies a 2-by-2 pixel square from a randomly chosen spot. Sample set 1 of Plate 1 (which can be found at the back of the book) shows a hundred examples of these 2-by-2 squares taken from a collection of fifty-nine photos. For comparison, sample set 2 shows a hundred 2-by-2 squares taken from a completely random image. The difference between the two sets is visually striking. As eye-catching as the random squares are, they clearly don't extend or blend colors the way the photographic squares do.[4] For example, about half of the photographic squares give the first impression of being one solid color, whereas none of the random squares do. Also, the 4 pixels making up a square are immediately discernible for only a few of the photographic squares, and in those cases the shade variations tend to be pleasantly subtle. For the random squares the opposite tends to be the case.

Notice the parallels between this comparison of pixel squares and our previous comparison of letter combinations. Just as we were able to spot incoherent letter combinations in small fragments taken from random typing (Figure 9.4), so too we're able to spot incoherent color blending in small fragments taken from a random image. In both cases coherence at this low level is a necessary start for building a fully coherent functional hierarchy of the kind represented in Figure 9.3, but it's a very meager start. Much more challenging levels of coherence must be built upon this low level if anything of significance is to come of it.

To see how hard it would be for a blind search to stumble upon coherence at any of the higher levels, all we have to do is build in the lower-level coherence. We didn't bother to demonstrate this for written instructions because the incoherence of those random seven-letter words convinced us that accidental word choice is as problematic as accidental letter choice. To do the demonstration for digital images, I used two of *Mathematica*'s[5] image-processing commands to transform the random image on the left side of Plate 2 into the one on the right. As a result of this processing, the new image has coherence not just at the bottom level of color extension and blending but also at the higher level of shape and boundary formation. That much coherence was supplied by the processing. Anything above that would be coincidental, and as we plainly see, there *isn't* anything above that—nor would there be if we were to spend the rest of our lives generating these images.

We can show that high-level functional coherence is hopelessly lost within the space of possible images by doing a calculation similar to the one we did for the space of possible keystroke combinations. If we say 1 in 20 squares from sample set 2 looks at least somewhat blended in color, as though it might have come from a photo, then the probability of a random image consisting only of such squares is calculated by starting with 1 and multiplying by 1/20 over and over, a total of 30,000 times—once for each of the squares that make up the full image. The resulting number is the fractional coverage of the image space by this very lenient target, so its reciprocal is—by the principle of reciprocal scale—the number of images a blind search would have to actualize to have a reasonable chance of finding that target. By now we know what this means. Were we to do the calculation, we would find this number to be so fantastically big as to make hitting even this uninteresting target a physical impossibility. Since the interesting target of all possible photos is much smaller, we know that it too is hopelessly outnumbered by images that would look completely random to us.

The Common Thread

For us to have made sense of such an odd assortment of inventions—digital photographs, a jury-rigged contraption for removing CO_2 from a space capsule, and half a page of written instructions—suggests we hit on something very general with the concept represented in Figure 9.3. Everyone who undertakes projects that require well-organized solutions should see something familiar in the hierarchical structure represented there—and that means everyone. Then again, so should everyone who marvels at living things.

Before being wowed with life in chapter 10, let's round out this chapter by completing our hike back down to the base of the mountain we just climbed. The simple sea-level theme connecting everything we've discussed is *the indispensable role of knowledge in the process of invention*. Starting in chapter 2 with tasks so simple we don't even associate them with invention—the making of an omelet or the wrapping of a present—we recognized the necessity of know-how even for these small accomplishments. And because they require know-how, our collective experience tells us they'll never happen unless someone who knows how *makes* them happen. This conviction we expressed as the universal design intuition: *Tasks that we would need knowledge to accomplish can be accomplished only by someone who has that knowledge.*

The value of our climb to the summit is that we now see why this intuition is correct. The making of an omelet is, in the terminology of chapter 6, the completion of a whole project. It is the bringing together of many small things and circumstances in just the right way to produce a big result. And we now see more precisely what we mean by this: those small things and circumstances must be arranged in a functionally coherent way, such that they all work together to produce something considerably more significant than the sum of the parts. Arrangements of this kind never happen by accident because they *can't* happen by accident. Making an omelet is easy for us not because it requires no skill but because we've mastered all the many simple skills it does require. And the very fact that each skill had to be mastered—from the cracking of eggs to the moving of objects in a coordinated way—shows that accidents are not likely to match those skills. Of all the things that *could* be tugged upon in the kitchen, the handle on the refrigerator door is only one. And of all the things that *could* be put in motion within the refrigerator—were it to be opened—the carton of eggs is only one. And of all the ways the egg carton *could* be moved, only a thin sliver of those possibilities is conducive to making an omelet. And so on. In the end, however ordinary the making of an omelet seems, the fact that a great many appropriate actions must be taken, each with a large number of ways to go wrong, means that fully appropriate *courses* of action are utterly lost within the staggeringly large space of raw possibilities. The advantage of know-how may seem modest at each tiny step, but after multiplying all these modest advantages, it becomes utterly decisive.

Knowledge is the primary ingredient of every omelet.

And if this is true for things as forgiving as omelets, then it's also true for the much more constrained things we call inventions. The fascinating conclusion from our mountaintop journey is that these special things—things that can only be made by the clever crafting of physical materials and actions, shaping them and combining them in just the right way to achieve a big result—trigger our design intuition for exactly the right reason. The reason we perceive purpose in inventions—busy wholes and whole projects—is precisely the reason they can't occur by accident: *they exhibit an organized functional coherence that can only come from deliberate, intelligent action.* They are conceived from the top down and constructed from the bottom up. They may *operate* by nothing more than physical causes, but they certainly don't *originate* that way. For them to be stumbled upon by coincidence, lost as they are within the vast space of raw possibilities, is simply not an outcome our universe can deliver.

Summing Up

In chapter 5 we asked why the tasks we need knowledge to accomplish are never accomplished without knowledge, and now we know why. Figure 9.5 summarizes the whole line of reasoning by which we found our answer. The core argument is simple enough to be stated in a single sentence:

> **Summary of the Argument**
>
> *Functional coherence makes accidental invention fantastically improbable and therefore physically impossible.*

The conclusion is summarized even more succinctly: *Invention can't happen by accident.* Invention requires know-how, and there is no substitute for know-how.

We did it! The dominoes have fallen. The conflict is over. Our design intuition has won!

Of course, our journey can't be replaced with two sentences. Use these summaries instead to gauge whether you've followed the argument in full and, if you have, as a way of quickly bringing to mind the main ideas. The reasoning isn't complicated, as scientific arguments go, but certain aspects will have been unfamiliar to many readers on first reading, which can be intimidating. Hang in there! If you had the determination to make it this far, you have what it takes to grasp the main points, with perhaps a second look at those you found challenging on first read. The figure legend will guide you to sections you may want to review.

In fact, even if you think the argument is wrong, I urge you to pause for a moment to make sure you've understood it correctly. I have more to say to persuade you, but this is the time to make sure the argument you disagree with is the one I'm making.

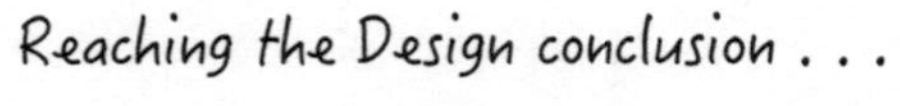

Figure 9.5 Two ways to conclude that inventions don't happen by accident. When we encounter even a very simple invention, like an origami crane, we automatically infer intentional design by recognizing that know-how was needed. This is the universal design intuition, depicted on the left. Having now carefully examined this inference, we see that it is fully affirmed by a series of sound deductions, which for the origami crane are given as six numbered statements. The general form of the reasoning is summarized in the single sentence on the right. For review, the universal design intuition is described in chapter 2; the concept of functional coherence begins to take shape in chapter 6 (see "Busy Wholes and Whole Projects"); chapter 7 shows that any inventive power in evolution must reside in repetition, not natural selection; chapter 8 examines the role of repetition in blind searches, showing that fantastic improbability means physical impossibility; and chapter 9 develops the idea of functional coherence fully, connecting it to invention and showing insight to be its only possible cause by showing accidental causes to be fantastically improbable.

CHAPTER 10

COMING ALIVE

The conflict within has been resolved. The tug-of-war between our design intuition and the consensus view of biological origins has been won by our intuition—*handily* so. As hoped, the win was not by the strength of technical science, though this certainly pulled for the winning side, but by the strength of *common* science—reasoning and observations we can trust because they're so closely connected to what we know from experience.

As significant as all this is, it hasn't yet given us a satisfactory answer to our big question: *To what or to whom do we owe our existence?* We have only a vague answer. We know we shouldn't wake up every morning thanking either natural selection or blind repetition for our lives. We know we weren't hatched by any egg-hunt search, which means we aren't the offspring of any accidental cause at all. This makes *purpose* a key ingredient of our origin—and perhaps many of us would be content to leave it at that.

Thomas Nagel has convinced me we should go further. As an atheist, he seeks to "explain the appearance of life, consciousness, reason, and knowledge neither as accidental side effects of the physical laws of nature nor as the result of intentional intervention in nature from without but as an unsurprising if not inevitable consequence of the order that governs the natural world from within."[1] Because our main accomplishment to this point has been to rule out the accidental cause that Nagel rejects, he would agree with us.

On the one hand, I view his agreement up to this point as a good thing, considering the quality of his thinking. On the other hand, because my purpose in this book has been to identify the source from which we came, I as a Christian thinker will feel as though I've come up short if an atheist thinker can be in agreement the whole way—even an exceptional atheist thinker like Nagel. We need to press on with the aim of reaching a more clear understanding of this nonaccidental source from which we came—an understanding that fits *either* Nagel's view of an impersonal power within nature *or* my view of a personal power outside nature, but not both.

These next four chapters of our journey will provide this understanding naturally without being overly forced to that end. We'll also address the most common reasons for doubting our conclusion that the design intuition has won. Both of these aims call for a closer look at life than we were ready for in chapter 6, benefiting now from our refined understanding of invention. This chapter will serve that purpose.

Taking Invention to a Whole New Level

However artfully humans have harnessed the regularities of the universe—fashioning the elements into things like smartphones and space telescopes—we can't escape the realization that someone has outdone us. The busy spider, the heroic salmon, the graceful orca, indeed *all* the living masterpieces that surround us demonstrate that physical materials and processes can be put to use much more elegantly than we ever have. Mind you, I say this as a lifelong technophile, not at all to disparage human invention but rather to remind us that life occupies a category that is unquestionably *above* human invention.

For example, among the more advanced products of human technology is a solar-powered underwater vehicle called *Tavros 2*. Operated by the University of South Florida, *Tavros 2* was designed to conduct monthlong missions in the Gulf of Mexico, measuring and reporting water depth and temperature. What makes this vehicle particularly sophisticated is that it operates autonomously, under the complete control of its onboard computer. *Tavros 2* is programmed to rise to the surface when it needs a solar recharge, after which it dives to its previous location and resumes data collection. If this aquatic robot had a résumé, GPS navigation would be listed under the *Technical Skills* heading, and tweeting would be under *Other Interests,* this being how it sends data back to the scientists at the marine lab (or to anyone else who likes to follow nerdy tweets).

But try comparing *Tavros 2* with something living. Dolphins, being of roughly the same size, might seem like a suitable species for comparison, but no sooner do we begin the exercise than we realize how incomparable these two inventions are. Like all robots, *Tavros 2* does exactly what it was programmed to do, whereas dolphins seem to do whatever they *want* to do. One is a physical machine while the other is, by all appearances, something profoundly greater than that—something *beyond* the mere physical.

In chapter 13 we'll explore the significance of this profound difference. For now, though, let's continue to concentrate on the physical aspects of living things—aspects that resemble machines, albeit machines of a most remarkable kind. We'll see that the machinery of life displays functional coherence on a scale that's presently beyond human comprehension, to say nothing of human imitation.

High-Tech Pond Scum

We'll start with a living "machine," by which I mean a form of life that, unlike dolphins, appears to operate in an entirely

Figure 10.1 *Tavros 2* being deployed by a University of South Florida researcher.

physical way. Far simpler than any of the individual cells within a dolphin is a lowly form of aquatic microbial life called *cyanobacteria*. Although cyanobacteria are single-celled organisms, the individuals of some species adhere to one another to form long filaments that interweave into huge mat-like colonies in stagnant or slow-moving water. They are, quite literally, the pond scum of the earth.

Despite occupying that humble position in the grand scheme of life, cyanobacteria are *light years* ahead of *Tavros 2* in terms of their technical sophistication. To see this, let's do some comparing and contrasting. One notable similarity is that *Tavros 2* and cyanobacteria are both solar powered. However, when we examine this feature in more detail, we find that the two aren't really comparable. The nonliving machine needs a solar collector the size of a coffee table, whereas the living one does very well with a collector roughly *one-trillionth* that size. And while the nonliving machine has only one trick for getting sunlight—surfacing—the living one is capable of much more. Filamentous cyanobacteria do control their depth in response to sunlight, but they're also able to coordinate complex sliding and oscillating movements to make their entire colony face toward the sunlight. So in terms of sophistication of movement for capturing sunlight, cyanobacteria have *Tavros 2* beat hands down (or maybe *filaments* down).[2]

The contrast becomes even more extreme when we consider manufacturing capabilities. *Tavros 2* has none, whereas every cyanobacterium houses an entire manufacturing plant within its microscopic walls. Powering all the operations of this plant is the process known as photosynthesis, which converts light energy into chemical energy. Much of this chemical energy is used to make sugar molecules from CO_2 and water, giving off oxygen (O_2) as a by-product. Sugar is therefore energy rich, which means that cells can "burn" it for calories. Alternatively, sugar serves as a versatile carbon compound cells can use to build the huge variety of other carbon-rich molecules needed for life.

Although we think of photosynthesis as a natural process, in the sense that it's happening all around us in nature, in another sense it is very *un*natural. More than any human invention, photosynthesis is an ingenious *exploitation* of the natural regularities of the universe, radically different from anything those regularities produce on their own. To grasp this, think of photosynthesis as the reverse of burning fuel, because that's what it amounts to. Burning is a very natural process, whereas *unburning* is not. With just a spark to get it going, oxygen readily consumes fuel molecules like sugars in its flames, forming CO_2 and gaseous water. By doing just the opposite, photosynthesis earns a position as one of those clever inventions, like air-conditioning, that harnesses natural regularities in order to work against them. And of these two inventions, photosynthesis is more clever by far.

The challenge for me is to give a sense of this without giving the equivalent of two or three chapters of textbook biochemistry. Thankfully, that can be done in much the same way that a wonderful book called *Stephen Biesty's Incredible Cross-Sections* gives us a sense of the engineering complexity of things like rescue helicopters and space shuttles.[3] Biesty artfully cuts the exterior away to show us how all the parts are arranged inside. Let's use a similar approach with *photosystem I*, one of the major components of the cyanobacterial photosynthetic apparatus.

The parts list for this engineering marvel (Figure 10.2) shows twelve protein parts and six smaller parts called *cofactors,* one of which (chlorophyll *a*) is used 288 times to build the full photosystem. These essential cofactors are held in their precise positions by the large protein framework, as shown in Figure 10.3.

The complete photosystem I shown at the bottom of Figure 10.3 has 417 pieces, each precisely positioned for the whole device to perform its function of gathering photons from the sun and converting their light energy into chemical energy. By my count, about three dozen genes in the cyanobacterial genome are dedicated to building this assembly: a dozen for encoding the protein components and two dozen more for encoding the enzymes needed to manufacture the cofactors. The whole assembly is massive in molecular terms, but with a diameter of just twenty-two billionths of a meter, *fifteen million* of these

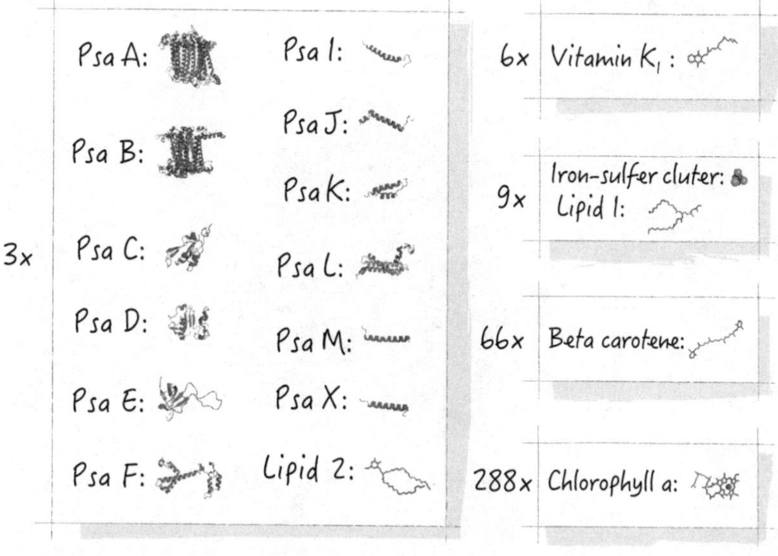

Psa A:
Psa B:
3x Psa C:
Psa D:
Psa E:
Psa F:
Psa I:
Psa J:
Psa K:
Psa L:
Psa M:
Psa X:
Lipid 2:
6x Vitamin K1 :
9x Iron-sulfer cluter: Lipid 1:
66x Beta carotene:
288x Chlorophyll a:

Figure 10.2 Parts list for building the cyanobacterial photosystem I.

things could fit in an area the size of a single pixel on an iPhone Retina display!

For those interested in learning about how the antenna system or the electron transfer chain (both shown in Figure 10.3) perform their respective functions within photosystem I, good resources are available online.[4] But you don't have to do any in-depth study to be fully convinced that photosystem I is an ingenious nanotechnological invention. All you have to do is let the diagrams speak for themselves.

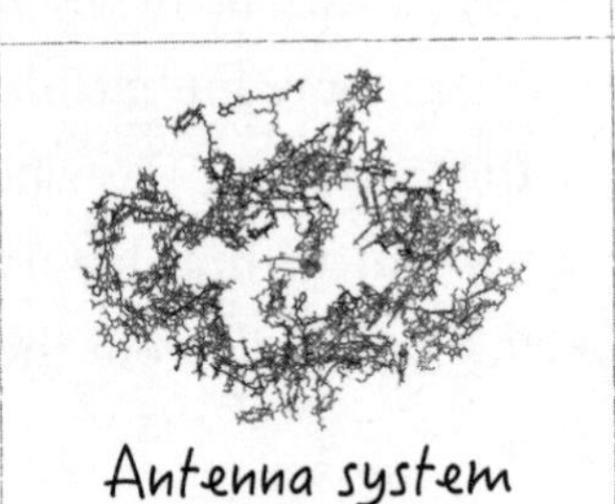

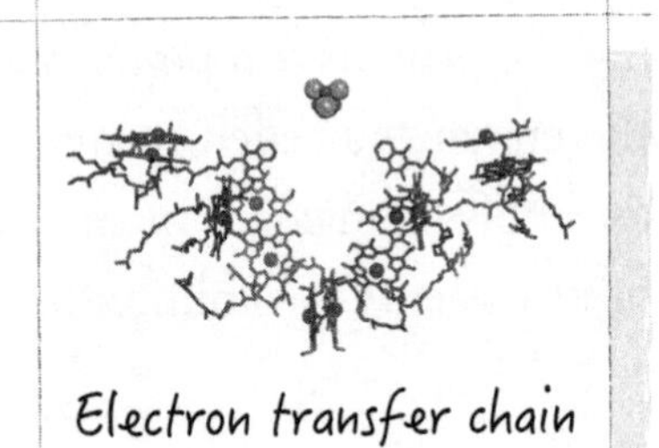

Figure 10.3 Cyanobacterial photosystem I with two of its important components shown above.

The very fact that the terms *electron transfer chain* and *antenna system* are used by the scientists who study photosystem I tells us that this photosystem's overall function involves multiple subfunctions, including the transfer of electrons and the collection of photons by an antenna. If you do decide to delve into the technical literature, you'll find a host of other functional descriptors, including *docking site, primary electron donor, initial electron acceptor,* and *quenching carotenoids.* Even if most of us have no clue what these terms mean, we all see that the high-level function of photosystem I depends on an extensive hierarchy of lower functions, and that should seem very familiar. This is another example of hierarchical functional coherence, made particularly striking by the tiny scale on which it has been implemented. As always, we immediately perceive this pattern to be a signature of purposeful invention.

As complex as photosystem I is, it's only one component of the many that make up the whole photosynthetic system. Figure 10.4 gives us an idea of how complex this whole system is. The figure is arranged in a hierarchical structure that should remind you of Figure 9.3. Topping the hierarchy is the cyanobacterial cell, shown as an actual cross-sectional image taken with an electron microscope. Below that is the photosynthetic system, which, though it is shown alone, is just one of many systems needed to support the top-level function of living life as a cyanobacterial cell.

At the next level down, the photosynthetic system is composed of two components: the thylakoid-membrane system and the CO_2-concentrating/reacting system. The first of these is responsible for harvesting light energy and converting this into chemical energy. The second is responsible for using this

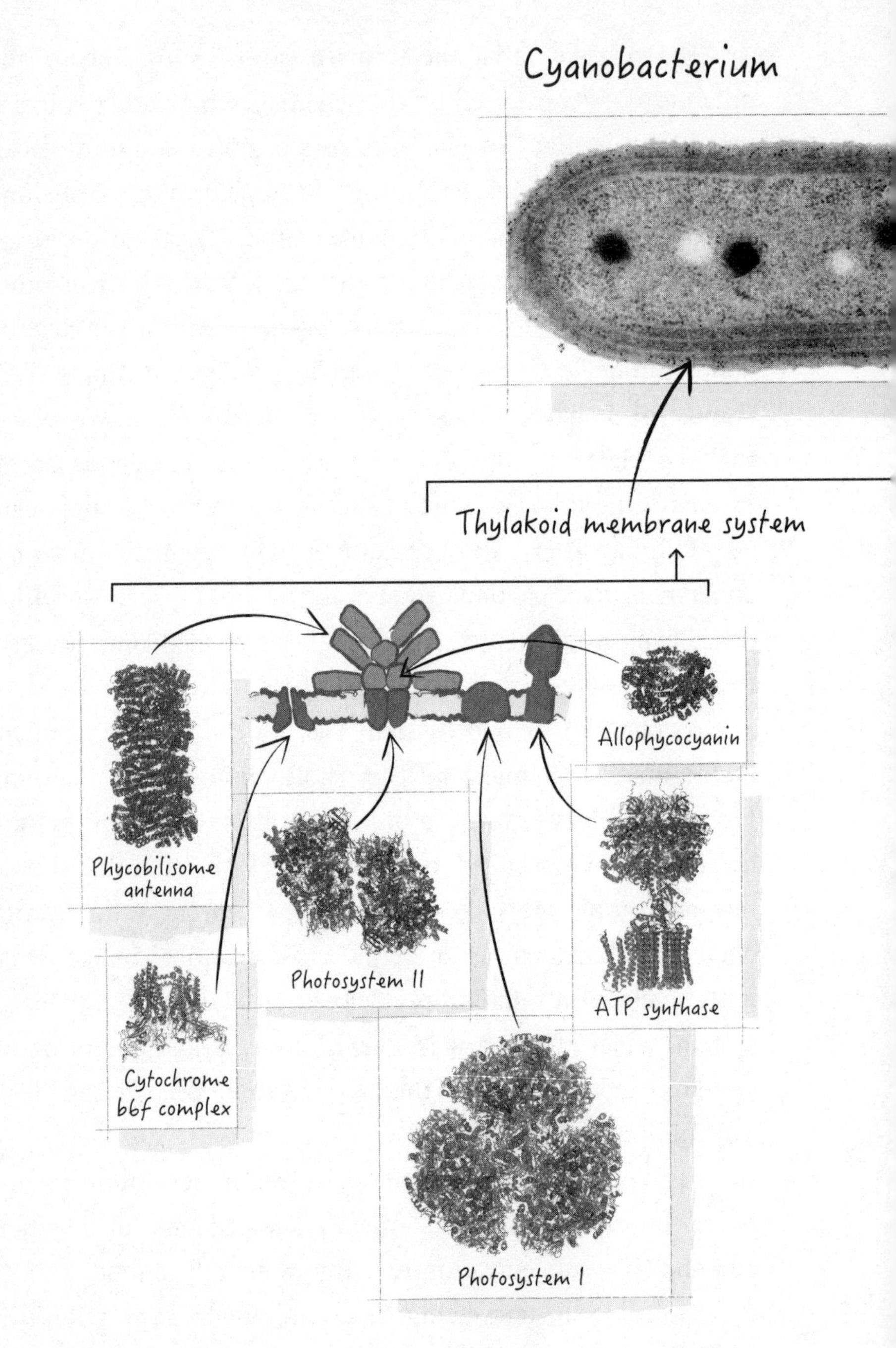

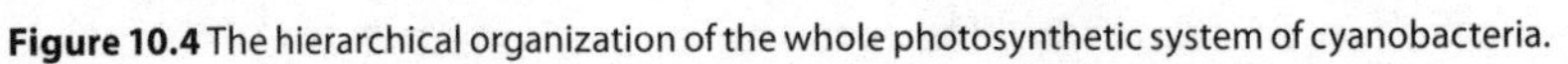
Figure 10.4 The hierarchical organization of the whole photosynthetic system of cyanobacteria.

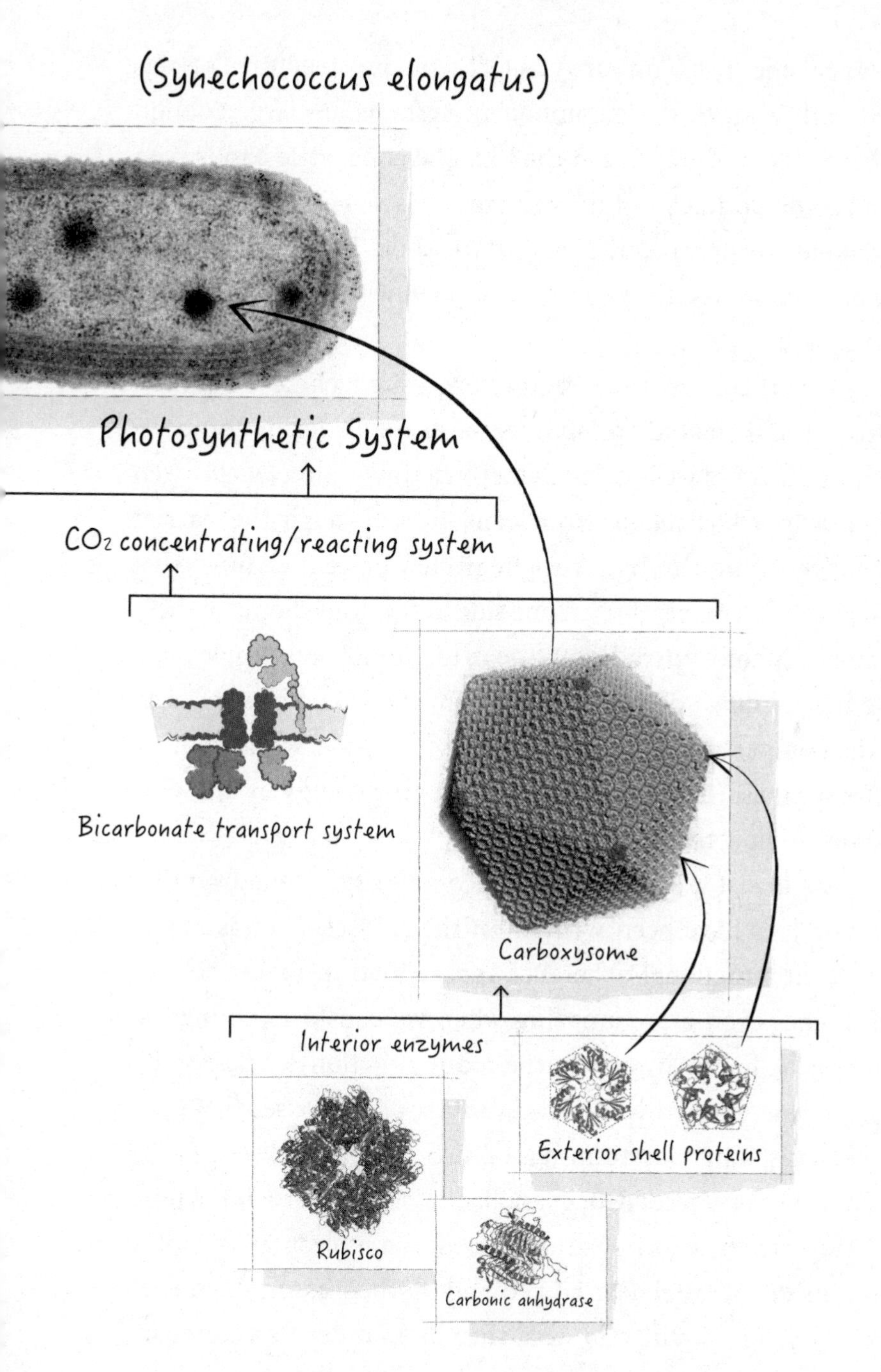
(Synechococcus elongatus)
Photosynthetic System
CO_2 concentrating/reacting system
Bicarbonate transport system
Carboxysome
Interior enzymes
Exterior shell proteins
Rubisco
Carbonic anhydrase

chemical energy to "unburn" CO_2. The major structures associated with both of these component systems are large enough to be visible in the image at the top. The concentric bands seen around the perimeter of the cell are the layers of light-catching thylakoid membrane. The large dark spots inside the cell are the carboxysomes, the reaction vessels in which the unburning takes place.

All of these functions require exquisite technical sophistication. The thylakoid membrane, for example, forms compartments that are so well sealed that even a tiny proton (a hydrogen atom stripped of its electron) can't pass through the barrier except by going through a sophisticated protein channel that systematically moves it from one side to the other. Some of these channels (photosystem II and the cytochrome b6f complex) act like tiny pumps, forcing protons from the "low pressure" side of the compartment to the "high pressure" side, while another (ATP synthase) acts as a turbine, extracting energy by allowing protons to flow the other way.

This is just a snapshot of the complexity of photosynthesis. Volumes have been written on the subject. And as amazing as the functional coherence represented in Figure 10.4 is, it becomes even more amazing when we consider the highest level of the hierarchy, where the many functions coalesce into one *purpose*. From this top-level vantage point, we see that photosynthesis, for all its stunning sophistication, is only one of the major functions needed for cyanobacteria to fulfill their purpose of being cyanobacteria. Ultimately, *all* the molecular assembly lines inside a cyanobacterial cell and *all* their associated genes and regulatory circuits do what they do in order for cyanobacteria to take their place among the spectacular living inventions that surround us—each so good that they cannot be other than what they are. When viewed through this lens, photosynthesis is one of those exquisite smaller inventions that serves its higher purpose so well as to make itself almost invisible.

And if the ability of cyanobacteria to make *sugar* from sunlight, air, and water has our eyes popping and jaws dropping, as indeed it should, try to imagine a proportionate response to the fact that they also make *cyanobacteria* out of those same raw natural ingredients![5] In fact, they make sugar only as a step toward making everything else—all the strikingly complex molecules that must be knit together into all the astoundingly complex systems and superstructures needed to form a living cyanobacterium.

It boggles the mind.

We're left to think that poor *Tavros 2* is really no more worthy of comparison to a lowly cyanobacterium than it is to an exalted dolphin. After all, raw natural ingredients like sand and metal ores and crude oil became *Tavros 2* only with the skillful help of thousands of people at hundreds of industrial plants of various kinds. With all due respect, this human invention does very little in comparison to the human effort expended to manufacture it. The contrast with cyanobacteria could hardly be more stark. With almost paradoxical genius, the inventor of these living marvels endowed them with the ability to manufacture cyanobacteria *by themselves*! That is, once the first cyanobacterium was made, however that happened, the rest were manufactured by cyanobacteria exactly as they are today—from air and sunlight and water.

Again, the mind boggles.

Coherence on Steroids

The cyanobacterium proves itself to be a dizzyingly impressive busy whole by accomplishing a dizzyingly impressive whole project—the manufacture of cyanobacteria—with apparent ease. And if that is so, then the toiling spider and the heroic salmon and the elegant orca can hardly be anything less. The sense of awe and wonder stirred in us by the humble cyanobacterium is only the beginning.

To leave you with a taste of the elegant complexity underlying familiar aspects of higher life, I have in Figure 10.5 traced one branch of the mammalian visual system from the top of its functional hierarchy down to the level of small molecules. Again, the point is merely to see the complex functional hierarchy that supports vision without having to understand it. However we choose to represent this hierarchy—however finely we divide the levels or the components that occupy each level—the hierarchy itself is very real and very impressive.

We've only scratched the surface, not just in breadth but also in depth. The truth is that living organisms are functionally coherent in a much more profound sense than human inventions are. Everything in an orca is completely and exquisitely devoted to the top-level purpose of being an orca. Every cell in the body both *sustains* the body and *is sustained by* the body. Living inventions are therefore *all-or-nothing* wholes—utterly committed to being what they are. The body is alive and thriving when all its parts are working, or it is dead and decaying when they are not. Apart from humans and the animals we tend to, nothing lingers very long between those polar extremes.

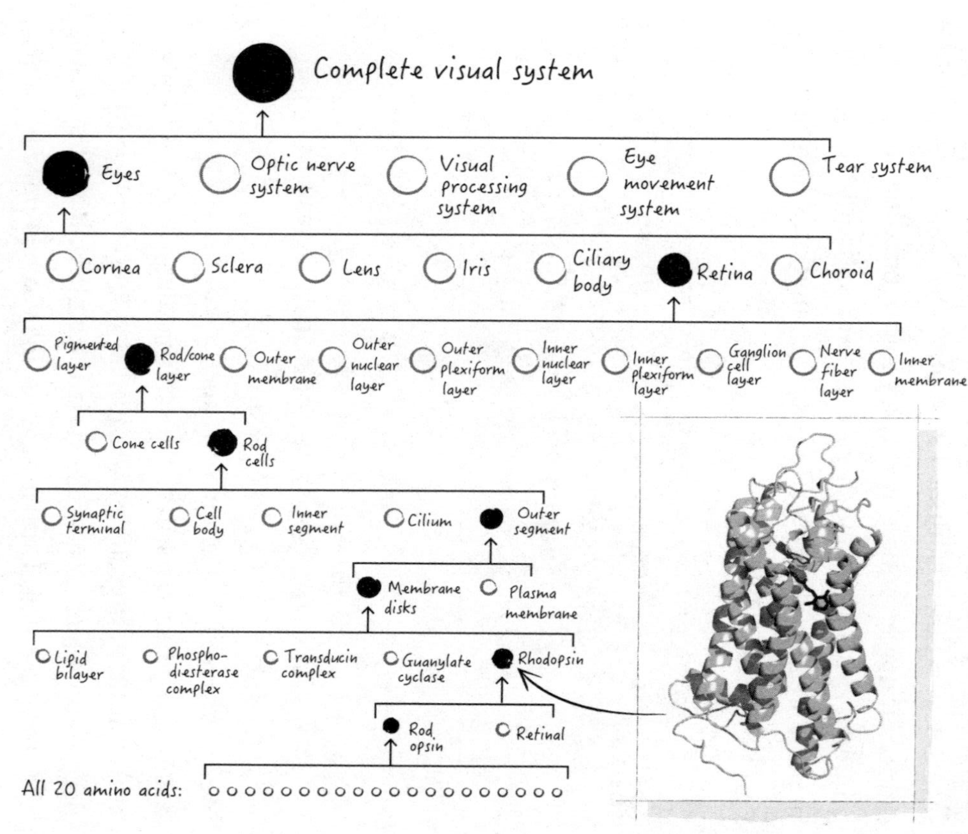

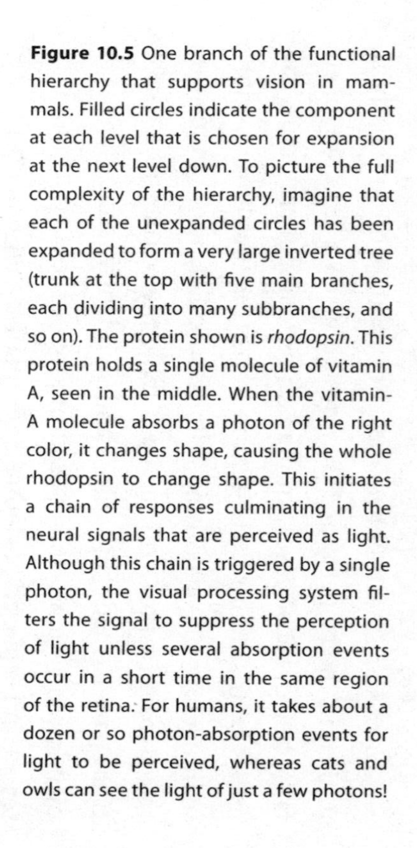

Figure 10.5 One branch of the functional hierarchy that supports vision in mammals. Filled circles indicate the component at each level that is chosen for expansion at the next level down. To picture the full complexity of the hierarchy, imagine that each of the unexpanded circles has been expanded to form a very large inverted tree (trunk at the top with five main branches, each dividing into many subbranches, and so on). The protein shown is *rhodopsin*. This protein holds a single molecule of vitamin A, seen in the middle. When the vitamin-A molecule absorbs a photon of the right color, it changes shape, causing the whole rhodopsin to change shape. This initiates a chain of responses culminating in the neural signals that are perceived as light. Although this chain is triggered by a single photon, the visual processing system filters the signal to suppress the perception of light unless several absorption events occur in a short time in the same region of the retina. For humans, it takes about a dozen or so photon-absorption events for light to be perceived, whereas cats and owls can see the light of just a few photons!

Cars and smartphones and pool robots are not nearly so unified in their operation. They do fail when a key component fails, but in most cases the remaining components are unaffected by that failure. The reason for this is that humans don't manufacture all-or-nothing wholes. Instead, we manufacture things part by part and then assemble the parts into a whole. Each part is made and tested independently, according to its own specifications, and indeed, many of these continue to be tested and replaced periodically even after they're incorporated into a whole. Life is nothing like that. Somehow, almost unbelievably, living inventions play a key role in building and maintaining *themselves*—all their parts formed and knitted together in unison *within* the whole. Life is never anything *but* whole.

The mind boggles.

> *Unlike human inventions, living inventions are all-or-nothing wholes. Every cell in the body both sustains the body and is sustained by the body. Life is never anything but whole.*

Making Sense with Amino Acids

We've seen from the laboratory experiments discussed in chapters 6 and 7 that Darwin's molecular fiddler is not at all adept at inventing new proteins, and in this chapter we've seen how deeply functional coherence runs through biological systems built from proteins. Until now, these may have seemed like separate problems for blind evolutionary searches: the problem of finding new proteins and the problem of finding helpful inventions that use proteins. In fact, the root problem in both cases is the impossibility of finding the necessary functional coherence by blind searches, because proteins, as molecular inventions, exhibit impressive functional coherence *in themselves.*

Figure 10.6 helps us understand what functional coherence means in the context of a single protein chain. The value of ribbon diagrams like the one shown on the left side of the figure is that we can see where the chain forms either of the two regular conformations that characterize all folded proteins: alpha helices (shown as coils) or beta strands (shown as arrows). But that visual clarity comes at the cost of oversimplification, as the more physically accurate stick representation on the right shows. Among the sticks we can, with some effort, discern a jagged version of the graceful path traced out by the ribbon on the left, but we also see what appears to be a messy jumble of darkly colored appendages jutting out from that path in all directions. Believe it or not, the functional coherence of this

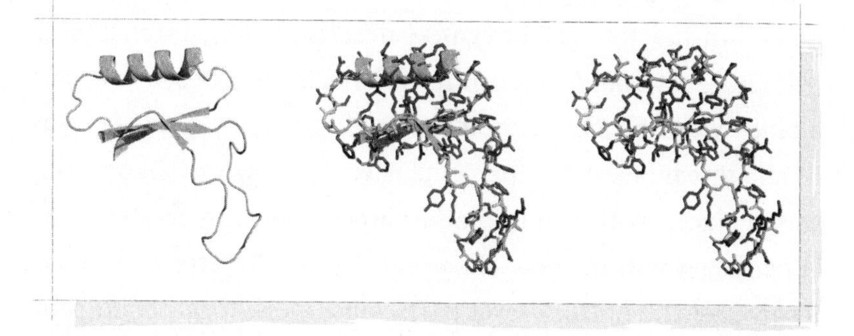

Figure 10.6 The role of the amino-acid appendages in forming protein structures. The three pictures each depict a portion of the smaller of two proteins that form the photosynthetic enzyme rubisco (shown at the bottom right of Figure 10.4). The middle image is a superposition of the ribbon diagram (*left*) and the stick representation (*right*).

protein lies within that complex "jumble," and the same goes for every other folded protein.

As we learned in chapter 3, the different appendage types are what distinguish the twenty amino acids. What looks like a mess to us is really an exquisite arrangement of amino-acid appendages along the whole protein that coaxes what would otherwise be a long floppy chain into forming a stable three-dimensional structure. Arranged sequentially in that special way, the appendages are more comfortable fitting snugly into their folded conformation than they would be flopping around wildly in the cellular fluid, the way a random sequence of amino acids does. Without their snugly folded conformations, the proteins of life couldn't perform their vital functions.[6]

Just how exquisite *are* the arrangements of amino acids that cause protein chains to fold, then? This is what I set out to measure with the experimental project I described at the start of chapter 5.[7] My aim in that work was to measure how improbable functional coherence is for these amino acids, in much the same way we assessed this for letters and pixels in chapter 9. I started by making lots of variants of the weakly functional penicillin-inactivating enzyme I described back in chapter 7 (the one that could be optimized by selection because it was already working as an actual enzyme). In each variant, a group of ten appendages that form a cluster, as shown in Figure 10.7, was replaced with random alternatives. You can think of the appendages within these clusters as being like letters or pixels in groups: the bottom-level parts must work together to produce something coherent. The idea was to assess this coherence by finding out how hard it is for a random assortment of appendages to be as functionally coherent as the appendages they replaced, meaning just coherent enough for the enzyme to work.

Once this was determined experimentally for the four clusters shown, the next step was to calculate the improbability of evolution stumbling upon that minimal functional coherence not just in those four clusters but in *all* the clusters needed for the protein to fold. I did this by converting the fraction of mutants that worked in each of the four experiments into an average probability of functional coherence per amino acid. I then multiplied this to estimate the likelihood of a fully randomized gene having the functional coherence needed to form a structure that supports enzyme function. As I said at the beginning of chapter 5, the result was striking. Of the possible genes encoding protein chains 153 amino acids in length, only about one in *a hundred trillion trillion trillion trillion trillion trillion* is expected to encode a chain that folds well enough to perform a biological function! So as hard as it was for our noise-seeking

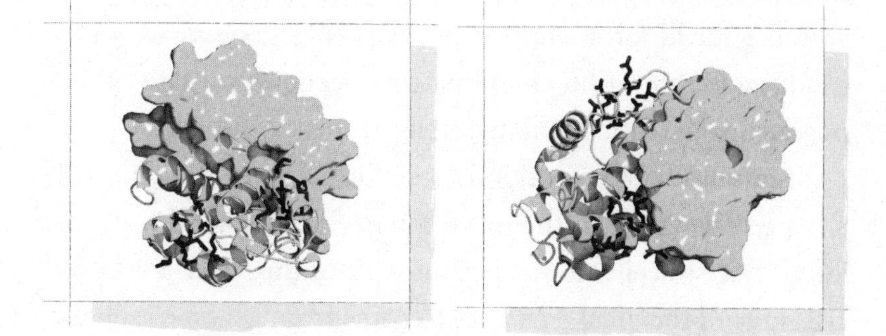

Figure 10.7 Two of the four clusters of ten amino-acid appendages that I randomized in order to measure the rarity of functional coherence are shown in the left picture; the other two are shown in the right picture. The beta-lactamase enzyme consists of a single protein chain 263 amino acids long. Its full structure divides visually into two portions called *domains,* which appear to be distinct units of folded structure. I focused on the larger of these two domains, which consists of about 153 amino acids. This one is shown as a ribbon diagram with randomized appendages shown as sticks. The other domain is shown in surface representation.

robot in chapter 7 to find a stadium, finding a biological invention is much harder, even at the low level of a single protein. We estimated that stadium noise may cover one part in a hundred thousand of the earth's surface, but the result here paints a much bleaker picture. Instead of the earth's surface for a search space, try picturing a sphere the size of the visible universe—*twenty-eight billion light years* in diameter—and instead of a target that covers six thousand square kilometers, try picturing one the size of a hydrogen atom! Now, *that's* a target we can safely write off as lost in space![8]

Invention Top-to-Bottom

As convinced as I am that protein folds are ingenious inventions in themselves, I don't want to give the impression that all of life's genius resides in proteins. Clearly it doesn't. As should be clear from Figures 10.4 and 10.5, the clever arrangement of amino acids to form working proteins is just one aspect of the exquisite design of life, and one that occupies a relatively low position in the functional hierarchy at that.

But while the invention of new life forms is undoubtedly a loftier exercise than the invention of new protein forms, that lofty exercise seems to require mastery of the lower-level exercise. One of the great surprises to come from genome sequencing projects is how many unique genes, and therefore proteins, are present in each form of life, including forms that to us look only subtly different. For example, a group of German scientists recently examined the genome sequences from sixteen cyanobacterial strains in an effort to discern all the distinct kinds of genes these strains carry.[9] Since they're all cyanobacteria, you might think they would carry the same set of genes, with perhaps an extra gene here or a missing gene there. The scientists found that they do share a common set of 660 genes, meaning not that these genes are identical from one strain to the next but rather that they are similar enough that we can be quite certain they encode proteins that fold to the same overall structure and perform the same biological function. Much more surprising, though, was their finding that nearly *14,000* genes are unique to individual strains! At an average of 869 unique genes per strain, this makes these bacterial strains more genetically different than alike, despite their overall external similarities.

The proportion of species-specific genes varies from one species to the next, but their existence in large number seems to be a property of *all* life, not just cyanobacteria. To quote the abstract of a recent technical paper, "Comparative genome analyses indicate that every taxonomic group so far studied contains 10–20% of genes that lack recognizable homologs in other species."[10] In other words, every species has many genes that seem, at first glance, to be one-offs—unlike any gene found anywhere else. The painstaking work of finding the structures of the proteins these genes encode is showing that about two-thirds turn out to resemble previously known proteins, with the remaining third being genuinely new.[11]

The origin of new categories of life does therefore seem to require the origin of new genes and proteins. Again, this isn't at all to say that the two are equivalent, but only that the one entails the other, with profound implications. Just as mastery of spelling and vocabulary is only the first step toward mastering writing, so too mastery of protein design is only a basic step toward mastering the design of life. The fact that mastery of this basic step is completely beyond the reach of blind evolution is therefore evolution's undoing. In the end, to believe the evolutionary story is to believe something much less plausible than hitting the *cosmic* cuna target—an atomic dot on a universe-size sphere—over and over in succession by blindly dropping subatomic pins.

No one should believe that.

The Fruits of Common Science

What we have deduced to be true of inventions generally—that they cannot happen by accident—is all the more true of the particularly remarkable inventions we see in life. What we realized at the end of the previous chapter—that omelets are completely lost within the space of kitchen possibilities—we can now extend to protein molecules within the space of amino-acid possibilities. And what is true for proteins is all the more true for the higher systems that use proteins for functions like photosynthesis and vision, and still more true for whole organisms occupying that highest of levels where the many functions coalesce into one purpose. Just as instructions and poems and love letters are completely absent from the mountains of QWERTY gibberish that can be accessed by blind searches, so it is with life. To do this activity we call *living* is so remarkable a feat that it can only be done by something extraordinarily well conceived and fashioned. Each and every new form of life must therefore be a masterful invention in its own right, embodying its own distinctive version of functional coherence at the very highest level.

I can only see these ingenious creeping, climbing, swimming, soaring, blooming, burrowing, luring, lunging, spinning, sporulating, fleeing, and fighting inventions as having come from the mind of God. To me, nothing else makes any sense.[12] That each one occupies its own unique place in our minds must surely reflect the fact that they were given their own places in the workshop of that supreme mind. There is no room for nonsense there—no thought of one masterpiece smearing into another, as if brilliant ideas could be blended like paint. That we have been chosen to behold the living wonders of that workshop ought to astound us, the keepers of our own workshops where we labor over much smaller projects. That we *came from* that workshop should astound us all the more. Among all the wonders that make Earth their home, we alone are compelled to stop and stare, to take this whole spectacle in—five parts inspiring to one part troubling—and to ponder it, knowing that none of it is accidental.

The children have been right all along.

CHAPTER 11

SEEING AND BELIEVING

The courage to defend our design intuition comes not just from the common-science argument we've developed but from the bigger picture as well. Everything seems to fit. Humans stand apart from all other living things as the one species that seeks wisdom and knowledge—the *sapient* species (*Homo sapiens*). If we knowers were meant to be, then surely we were meant to *know* we were meant to be. And indeed we do. Well before our formal education begins, we have already mastered the simple science of interpreting our everyday experiences. This science produces in our young minds the universal design intuition. With or without parental approval, we know instinctively that living wonders so remarkably good at being what they are—spiders being spiders and orcas being orcas—exist only because someone made them for the express *purpose* of being what they are. If you saw this instinct as being more heart than head before you started reading, I hope our journey has corrected that imbalance.

Does anything *not* fit, then? This is an important question to ask whenever we think we've come to a correct understanding of a contentious subject. It's not a question of *completeness* but rather a question of *contradiction*. Indeed, as we'll see in the final chapter, acknowledging that science shows life to be designed hardly begins to answer the important scientific questions. It merely opens the door to a correct conception of biology—a door that has been blocked and barred for well over a century. The weighty intellectual challenge of building that long-awaited correct conception—after thinkers have filed through this door in large number—has barely begun. And that's perfectly fine.

The first aim for this chapter is simply to consider whether we've overlooked any facts that somehow refuse to fit into this otherwise coherent picture of a designed world. If we haven't, my next aim will be to equip experts in common science—like you—to stand firm in a world where certain experts in technical science do their best to push others around.

To begin, we look to those who've been working under the flag of materialism, which (unsurprisingly) is also the flag of Darwinism.

The View from the Stands

According to journalist Paul Rosenberg, writing for *Salon*, "things could get a whole lot worse for creationists because of Jeremy England, a young MIT professor who's proposed a theory, based in thermodynamics, showing that the emergence of life was not accidental, but necessary."[1] By *necessary,* Rosenberg means so physically inevitable as to be unremarkable. England does seem to espouse this no-big-deal view of life's origin. "You start with a random clump of atoms," he says, "and if you shine light on it for long enough, it should not be so surprising that you get a plant."[2] Rain happens. Life happens.

What are we to make of this? In particular, what should you do if you feel certain this MIT professor is wrong but also know you'll never be able to follow his argument? You could search the web to find people with Ph.D.s who dispute his claim, but there would probably be people with Ph.D.s who dispute the disputers as well. And so on. In the end, this technical to-ing and fro-ing gives little aid to nonexperts, apart from the comfort of knowing that at least some experts are on their side.

However, if the decisive matters in this discussion belong not to the technical disciplines but rather to common sense and common science, as I've claimed, then this picture of nonscientists as spectators at a sporting event—where most players are wearing the Darwin jersey—is all wrong. When it comes to simple intuitive reasoning, the playing field is level, and everyone is qualified to play.

The Inconceivability of Accidental Invention

As common scientists move down from the stands and flood the field, the most important advice for them to bear in mind is the familiar call to "keep your eye on the ball." We have arrived at what looks to be a decisive argument. In a sentence: *Functional coherence makes accidental invention fantastically improbable and therefore physically impossible.* Invention can't happen by accident. This is the ball, then. To become distracted by any defense of accidental origins that doesn't answer this argument is to take our eye off the ball. Instead, we're wondering whether there's a single piece of work out there that should convince us this argument is wrong.

What would this even look like? Can we be wrong to attribute functional coherence to biological systems? I can imagine people thinking this is wrong, but only out of ignorance. Certainly ignorance as to the necessity of functional coherence within cells existed among some biologists of Darwin's day. Writing in 1868, nine years after the publication of *On the Origin of Species,* German biologist Ernst Haeckel said the following about aquatic microorganisms he classified under the heading *Monera:*

> These very simplest of all organisms yet known, and which, at the same time, are the simplest imaginable organisms, are the Monera living in water; they are very small living corpuscles, which, strictly speaking, do not at all deserve the name of organism. For the designation "organism," applied to living creatures, rests upon the idea that every living natural body is composed of organs, of various parts, which fit into one another and work together (as do the different parts of an artificial machine), in order to produce the action of the whole. During late years we have become acquainted with Monera, organisms which are, in fact, not composed of any organs at all, but consist entirely of shapeless, simple, homogeneous matter. The entire body of one of these Monera, during life, is nothing more than a shapeless, mobile, little lump of mucus or slime, consisting of an albuminous combination of carbon. Simpler or more imperfect organisms we cannot possibly conceive.[3]

As you may have guessed, cyanobacteria—the stunningly sophisticated photosynthetic marvels we encountered in chapter 10—are among the bacterial species to which Haeckel referred here. He couldn't have been more wrong about their internal structure, and moreover, his error can't be excused as though no one knew better back then. Some two hundred years earlier Antonie van Leeuwenhoek, one of the pioneers of light microscopy and the father of microbiology, observed the complex powered movement of many bacterial species in water.[4] Add to this the observation of bacterial cell division and the conclusive demonstration by Louis Pasteur that bacteria only come from bacteria—all well in place before 1868—and there's really no excuse for Haeckel to have missed the fact that remarkable processes were going on inside these little creatures. Indeed, the tiny scale of those processes should have brought recognition that they had to be far more sophisticated than the artificial mechanisms he mentioned—clocks and steam engines and the like.

Despite his blunder, the quote shows that Haeckel had a well-formed notion of functional coherence, evident in his description of a hierarchy of components working together to form a functional whole. What he lacked was the conviction that sophisticated functions are never achieved *without* functional coherence. No one with an interest in biology makes this mistake today. That living things all the way down to bacteria are chock-full of systems that exhibit functional coherence all the way down to their molecular constituents is now such a pervasive theme in biology as to be unmissable.

As for the connection between functional coherence and fantastic improbability, here again we have something that can be overlooked but not refuted. Interestingly, even one of the most ardent defenders of Darwinism in recent times, Richard Dawkins, has not overlooked it. The first chapter of his 1986 book *The Blind Watchmaker* is titled "Explaining the Very Improbable." There he describes the connection as follows:

> *However many ways there may be of being alive, it is certain that there are vastly more ways of being dead,* or rather not alive. You may throw cells together at random, over and over again for a billion years, and not once will you get a conglomeration that flies or swims or burrows or runs, or does *anything,* even badly, that could remotely be construed as working to keep itself alive.[5]

The very same principle applies at levels above and below the cell. Coherent skeletons are impossibly rare among random arrangements of bones, as are coherent body plans among random arrangements of organs, and molecular machines among random arrangements of folded proteins, and folded proteins among random arrangements of amino acids. According to our analysis, *none* of these inventions had any prospect of coming together by accident. They all required insight.

Dawkins still thinks natural selection can do the work of insight, but we know better. Interestingly, his own words point to the gaping hole in Darwin's theory, which we saw back in chapter 7. Natural selection happens only *after* cells are arranged in ways that work to keep the organism alive, so selection can hardly be the *cause* of these remarkable arrangements. Darwin's simplistic explanation has failed, and the millions who have followed him have nothing but his outdated assumption to stand on.

The stepping stones over which these followers think life has skittered from one form to the next are definitely not explained by natural selection. Selection steps to forms that already exist, so it doesn't explain the forms themselves, much less the intricately engineered circumstances that would have been needed for these forms to be connected through lines of descent. And the problem never goes away. Because the impossibility of accidental invention is at the root, and because each new form of life amounts to a new high-level invention, the origin of the thousandth new life form is no more explicable in Darwinian terms than the origin of the first. Even if we suppose a first insect to have been formed somehow—without trying to explain how—all the countless insects that differ substantially from that first one would still be new top-level inventions. The component inventions common to all insects would have had their specific representations in that first insect, but a great many of these components would have had to be substantially reworked to suit each new insect. This would have been a staggering feat of re-engineering in itself, to say nothing of the great variety of *new* components that would have had to be invented from scratch. In the end, each new form of life amounts to a stunning new invention, and since the hallmark of invention is functional coherence—which accidental causes can't explain—we rightly see each form as a distinct masterpiece.

Accident is out of the picture. Stepping stones connecting these masterpieces are either a figment of our storyteller imaginations or proof that God has, at times, converted the world into an exquisite nanofabrication facility. There is no substitute for brilliance, so either the stones are *part* of the brilliance or they aren't anything at all. The genius of life is not in question. The only question is how The Genius of life did his work.

> *Because each new form of life amounts to a new high-level invention, the origin of the thousandth new life form is no more explicable in Darwinian terms than the origin of the first.*

Touring England

Returning to England—*Jeremy* England—and my aim of liberating readers from their dependence on experts, I don't mean to suggest that nonexperts should ignore the debate among experts. The reward for following that debate, even as a casual observer, is a sense of how things are shifting within the academy, which is worth having. So while I hope every reader is able to say why England's equation

light + random atoms + time = living plant

can't be correct, I think readers will also be interested to know how one of the world's leading chemists views this idea of life originating by chemical accident.

I'm referring to Jim Tour, professor of chemistry and nanoengineering at Rice University, whom I met after a stunning presentation he gave at a meeting at Baylor University in 2009. The best way I can describe his work is to say that he and his team do with atoms what kids do with construction toys. If you think I'm kidding, try googling *nanocar* or *nanodragster*.

When it comes to understanding, from firsthand experience, the difficulty of making atoms come together to form molecular devices, very few people can match Jim Tour. I certainly can't, and I'm pretty sure Jeremy England can't either. With all due respect to England and his theory, then, it would be interesting to know what Tour thinks about the casual confidence so many scientists seem to have in the ability of unguided natural processes to build complex molecular devices.

Thankfully, we don't have to wonder about this. Speaking of the separation of helpful products from unhelpful ones after each step in a complex synthesis procedure (without which the procedure would fail), Tour says:

> If one asks the molecularly uninformed how nature devises reactions with such high purity, the answer is often, "Nature selects for that." But what does that mean to a synthetic chemist? What does selection mean? To select, it must still rid itself of all the material that it did not select. And from where did all the needed starting material come? And how does it know what to select when the utility is not assessed until many steps later? The details are stupefying and the petty comments demonstrate the sophomoric understanding of the untrained.[6]

In other words, the only thing people demonstrate when they assume such things can happen by accident is that they don't know what they're talking about.

The Magician's Hat

For those occasions when you don't have someone like Tour at your side, here's a simple way you can test supposed proofs that accidental invention works. Think of the illusion of pulling a rabbit out of an empty hat. What makes this trick entertaining is that we seem to be witnessing the impossible. We know a rabbit can't come out of a hat unless it first went in, and yet we have the impression that nothing went in except the hand now holding the rabbit. It looks like magic in terms of this immediate impression, but a broader perspective assures us it's merely an illusion, even if we have no idea how the trick was performed. After all, if anyone really had the ability to bring things into existence out of nothing, they would find a more productive way to use their superpower than by working as an entertainer.

Both the impression of magic and our ability to analyze that impression in this way—by surveying the bigger picture—will help us know what to make of supposed demonstrations of the power of evolution. Think of the hat as a conceptual black box that surrounds and conceals all the inner workings of one of these demonstrations. As with the rabbit trick, our strategy is to compare what went in with what came out, without worrying about what happened inside. In doing this, we should pay particular attention to *knowledge* because of its essential role in invention (Figure 11.1).

The first question to ask of a demonstration is whether it even gives the *appearance* that the impossible has occurred. If not, then it clearly doesn't address our argument. Our claim is very simple. Having noticed that we intuitively know invention can't happen by accident, we believe we've now come to a firm understanding of why this intuition must be correct.

What went in?	What came out?	Must be magic?
A hand	A hand holding a rabbit	*Yes!*
A rabbit and a hand	A hand holding a rabbit	No: Arranged by magician
Alphabet soup and heat	Heat and instructions in alphabet soup	*Yes!*
Alphabet soup and heat	Heat and alphabet soup	No: No arrangement
Light and random atoms	A plant	*Yes!*
A physicist	A physicist with a theory about the origin of life	No: Arranged by physicist
A lifeless planet	A planet teeming with life	*Yes!*
God and a lifeless planet	God and a planet teeming with life	No: Arranged by God

Figure 11.1 Using the magician's hat to test demonstrations. The beauty of this approach is that you don't have to know what happens inside the hat. You simply ask whether what came out—the final effect or outcome—can be explained in terms of what went in. Examples are given in pairs to show how a correct result requires correct identification of the inputs and outputs. Notice that only if what happened required knowledge and no knower was present do we conclude that magic has occurred.

To counter this claim, someone would have to show that what both intuition and calculation affirm to be impossible somehow *isn't* impossible. Anyone not even pretending to do this hasn't understood what needs to be demonstrated.

Of the many demonstrations I've encountered over the last thirty years, not one passes this test of relevance. No one has said, "Look! We've found a way for the impossible to happen!" Instead, they offer unsurprising examples where searches that should succeed do, or where selective homing that should work does. In doing so, they ignore the fact that invention by accident requires *fantastically improbable* searches to succeed. Since *that* is the unbelievable claim, *that* is what they would have to demonstrate. And if they did? Well, their demonstration would be the world's first scientifically proved, mathematically validated instance of magic. Even then, I think we would find it viscerally impossible not to attribute the outcome to an invisible knower, which would leave our design intuition intact.

Passing the Hat (or Not)

A couple of examples will prepare you to use the hat yourself. The first is a demonstration that Richard Dawkins offered in *The Blind Watchmaker*, where a computer program started with a random sequence of twenty-eight letters and spaces and ended up with the Shakespearean line **`METHINKS IT IS LIKE A WEASEL`**, supposedly by evolution. Let's ignore how the program worked for a moment in order to see how the hat works. In this case, since Dawkins was as much a part of the demonstration as his computer, both he and his computer went into the hat. After some time (how long is of no importance), he came out with **METHINKS IT IS LIKE A WEASEL** displayed on his computer. Now, on the face of it, should anything about this amaze us? Clearly not. For a person with a computer to produce a written sentence is nothing out of the ordinary. Sentences magically appearing in oracle soup would amaze us (to put it mildly), but this is nothing of the kind.

That was easy.

Now that the hat did its job, a quick peek inside will be informative. Dawkins designed his program to carry out two simple steps repeatedly. The first step was to produce lots of copies of the parent sequence, starting with the random one, with occasional random typos in them. In the second step, each copy was compared to the target sentence **METHINKS IT IS LIKE A WEASEL**, and the copy with the most correct letters, however few, was selected as the parent for making a new batch of copies, and so on. After about forty rounds of this, an exact match was found.

Dawkins knew this wasn't blind evolution, of course. His intended point was simply that cumulative selection, where improvements are allowed to build a little bit at a time, can accomplish what would never be accomplished if the whole finished thing had to appear at once. In his words, "If . . . there was a way in which the necessary conditions for *cumulative* selection could have been set up by the blind forces of nature, strange and wonderful might have been the consequences."[7] Granted. But then strange and wonderful assumptions *often* imply strange and wonderful consequences, don't they?

Once again, what's envisioned here is an extensive network of natural stepping stones that happen to line up in ways that make selection take extraordinarily insightful paths. We've already exposed this ploy. Accidental stepping stones leading to these fantastically improbable destinations would *themselves* be fantastically improbable. If we needed further proof that such remarkable coincidences never happen by accident, we can thank Dawkins for supplying it. It isn't hard to imagine some practical need calling for a line of Shakespeare—a homework assignment perhaps. But when we bring this to mind, we instantly see that the following line of gibberish (presented by Dawkins as the first selected sequence) wouldn't meet that need:

WDLTMNLT DTJBSWIRZREZLMQCO P.

Equally unintelligible sequences meet other needs, of course—long passwords or encrypted messages. What we can't imagine, though, is an honest-to-goodness series of these unrelated needs just *happening* to line up in such a way that they connect Dawkins's original random sequence to **METHINKS IT IS LIKE A WEASEL**. That *definitely* won't happen by accident, which is why Dawkins had to line up the stepping stones himself. Somehow, though, he thinks the implausibly complex network of stepping stones that would be needed for life to evolve *did* line up by accident. And somehow he thinks his thoroughly unremarkable demonstration should convince us of that thoroughly unbelievable claim.

We know better. Natural stepping stones may lead to strange and wonderful destinations in our imaginations, but the real world is different. Nothing becomes useful or wonderful until functional coherence is present in good measure, and whatever helpful things the natural world may supply in good measure, functional coherence isn't among them.

Nearly twenty years after Dawkins's demonstration came another worth mentioning, this one announced on the cover of *Discover* magazine with the words *Testing Darwin—Scientists at Michigan State Prove Evolution Works.*[8] What supposedly evolved was a computational function, so you would need a bit of technical knowledge to understand what came out of the hat. In just a moment I'll show how the hat comes through even without this knowledge, but first let me give you this assurance: the computational function that was produced was so elementary that it wouldn't have merited attention *apart from* the claim that it evolved.[9] So since computer science was one of the competencies the scientists brought to the project, we again have a situation where what came out of the hat is not the least bit remarkable considering what went in.

Still, one aspect of this demonstration may seem to challenge our conclusion about functional coherence, at least at first glance. The output function in this case required nineteen or so elementary machine instructions to be arranged into a working whole, and the investigators didn't explicitly supply this arrangement, the way Dawkins supplied his. This seems to imply that functional coherence was produced over the course of this evolutionary simulation.

What are we to make of this? First, keep in mind that our claim is not that blind processes are incapable of producing any functional coherence at all but rather that they are incapable of producing it in the amounts needed for useful *inventions.* We've already seen very small amounts of functional coherence appear by chance, as when the word *ink* appeared in half a page of random typing or when a random grouping of four pixels just happened to have blended colors. And this can be edged upward a bit by sifting through randomness on a larger scale. Just now, I wrote a program to do this with random typing and found the longer word *bobbled,* which involves nine coherent keystrokes, including the spaces before and after. But the increase in coherence came at a considerable cost, as is always the case for blind searches. To find this solitary seven-letter word, my program had to sift through more than *fourteen thousand pages* of nonsense!

Serious invention requires not just a smidgen of functional coherence but extensive amounts arranged over a hierarchy of levels, and this simply can't happen by accident—for *any* kind of invention. The action of bulldozers moving junk heaps at the dump, for example, may well cause a ball bearing to find a makeshift socket or a lever to find a crude fulcrum or a cable to wrap itself around a cylinder, but none of these simple arrangements do anything significant enough to rise above the junk. Not even on a trillion trillion planets *covered* with junk would an accidental robot ever rise up and flee from the bulldozers, much less scurry around looking for parts to build a copy of itself.

Once this hard fact is grasped, the thought of quibbling over whether nine coherent keystrokes or nineteen coherent machine instructions ought to be heralded as significant inventions becomes pointless. Both are completely insignificant compared to what people commonly set out to accomplish with words or with computer code, to say nothing of all these extraordinary accomplishments we call life.

There's more to this story, though, at least for those able to dig deeper. If you have the ability to dissect demonstrations that "prove evolution works," you'll find that researchers commonly embed their knowledge of what was needed for success into their evolutionary models. In other words, there is *cheating* going on here, though the researchers may not think of it as such. In a way, it's hard *not* to cheat with these simple demonstrations. The problem is that the researchers know too much. In particular, they know how the search should be conducted if it's to have any chance of succeeding, and because they *want* it to succeed, it's nearly impossible for them to avoid helping it along.

For example, the scientists who reported the evolution of the computational function had to offset the cost of useless genetic instructions in their digital "organisms" by rewarding them in proportion to the size of their genomes. As we saw with the stepping-stone experiment at the end of chapter 7, real life behaves very differently. Genes that don't work are a burden, and nature has no incentive program to offset this burden. The scientists who did the computational project knew this but used a very unnatural version of selection anyway, just to get the outcome they wanted. Additional instances of guidance have been documented for that study and for several other demonstrations claiming to show that evolution works.[10]

Most of us can't dig that deeply, though. In fact, if we don't even understand what came out of the hat, how are we supposed to decide whether it looks like magic? As I said, the hat comes through even here. Instead of asking whether the demonstration looks like magic to you, ask yourself whether it seems to look like magic to the people who understand it. Are *they* acting as though they've encountered a fountain of invention? Are the experts trembling with astonishment? Are investors scrambling to get a piece of the action? Are technology companies letting all their smart people go, convinced that human insight has now become superfluous? Or is the response perhaps more muted?

Look in particular at the scientists who conducted the demonstration. Are they hanging on to their day jobs—just as magicians do? If so, this is a sure sign they didn't *really* pull a rabbit out of an empty hat.

Fiddler for Hire

As a finder of inventions, Darwin's evolutionary mechanism is a complete bust, but as we saw in chapter 7, it sometimes comes in handy as a fiddler. The example I described where a weakly functional enzyme was dramatically improved demonstrated this. The starting point was ideal: a complete invention, having all the necessary components in place and working, but not fine-tuned for optimal performance. Fine-tuning involves the adjustment of many small details, so trial and error is often the best way to do it. In these situations, *selective optimization*—a tuning process that repeatedly selects the best variant after introducing slight variations—is often useful.

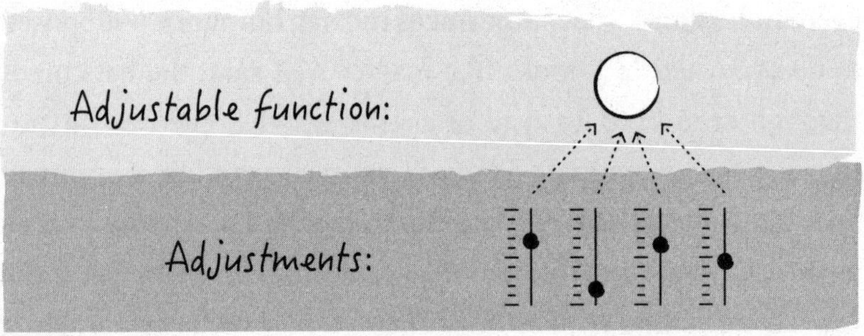

Figure 11.2 The relationship between adjustments and the function they affect. Dotted arrows indicate that the function being adjusted is not *caused* by the adjustments, in contrast with solid arrows in Figure 9.3. Still, it's common for adjustments to influence the complex arrangement of parts that *is* the cause (not depicted here) very significantly.

A picture may help us see that selective optimization, while useful, is nothing like invention. The contrast between Figure 11.2 and Figure 9.3 illustrates the differences. Most importantly, as the experiments described in chapter 7 showed, selective optimization only works on a preexisting function. Unlike the lower-level functions of Figure 9.3, the adjustments depicted in Figure 11.2 don't *cause* the high-level function; they merely *tune* it. This high-level function wouldn't be there to be tuned, apart from a functionally coherent arrangement of components to cause it.

For example, NASA has used selective optimization to help design some of its antennas, including the small one shown in Figure 11.3. This antenna, which looks like nothing more than a randomly bent paper clip mounted on a threaded base, actually has its bends in just the right places to enable it to work well. Antenna shapes are ideally suited to selective optimization because nearly every shape works to some degree, and yet small

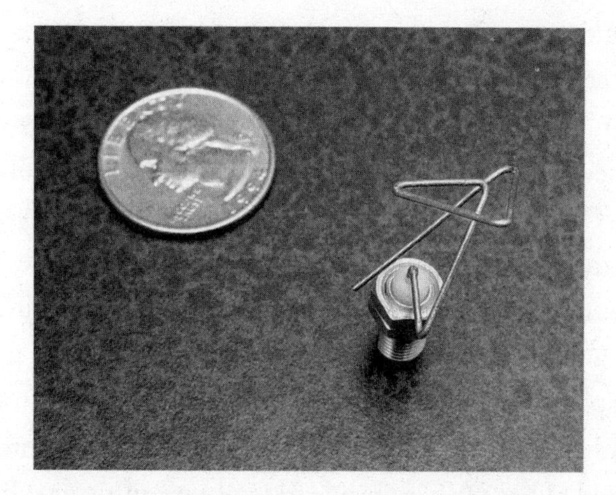

Figure 11.3 One of the small antennas designed by NASA with the help of computational selective optimization (see www.nasa.gov/centers/ames/news/releases/2004/antenna/antenna.html).

adjustments have measurable effects. Notice that this kind of optimization is really an *application* of human insight rather than a *replacement* of human insight. No computer will ever design an antenna by accident. Instead, human engineers who know all about antenna design must carefully set the stage for computerized optimization to do exactly what they intend for it to do. In other words, *humans* invented this little antenna by bringing together everything needed for its design, including the computational search that refined it.

This means the antenna's origin fails the hat test: it looks nothing like magic because the needed understanding was supplied in the usual way. But this realization doesn't detract from the value of the antenna or the value of selective optimization. Both are clearly useful. So keep in mind that the hat test isn't a test of usefulness or scientific validity but a test of relevance to our argument against accidental invention. By failing the hat test, the antenna merely shows that it's not an invention that appeared by accident. It *is* an invention, of course—just not an accidental one.

A team of scientists from Cornell and the University of Wyoming recently presented a more entertaining example, reminiscent of something from a Pixar film.[11] Think of pulsating Jell-O cubes sticking together to form jiggling bodies that flee from spoon-wielding children, and you get the picture (Figure 11.4). Like Pixar inventions, these Jell-O creatures exist only in a computer-generated world, but unlike their film-star counterparts, the Jell-O creatures had to "learn" how to run by trial and error. Taking hints from biology, their scientist creators used cubes of three kinds to facilitate running. One kind actively contracts and expands in a rhythmic way, mirroring the rhythmic contraction and stretching of leg muscles during running. Cubes of the other two kinds have passive structural roles, one rigid like bone and the other more flexible, like cartilage.

Because the pulsing cubes all pulse to the same beat—some contracting on odd beats and others on even—you'd think that even a random mass of cubes would have a reasonable chance of jiggling itself in one direction, as long as it contains a good number of pulsers. Perhaps it would, but more effective motion requires nonrandom organization of cube types into extended regions, as indicated by the light and dark coloring in Figure 11.4. Taking more cues from biology, the inventors imposed rules for these nonrandom regions. Selective optimization was then used to refine runners within the imposed constraints. Like the NASA antenna, these Jell-O runners show how selective optimization can be cleverly exploited within the context of a large project that's conceived and implemented by humans. But as useful as this is, it has the familiar look of a tool in the hands of inventive humans—one more technique among countless others that we humans have discovered and refined to suit our inventive ends.

Interestingly, humans weren't the first to use this tool. Selective optimization finds elegant application in life, the most notable example being the process of antibody refinement (known to biologists as *affinity maturation*), which plays

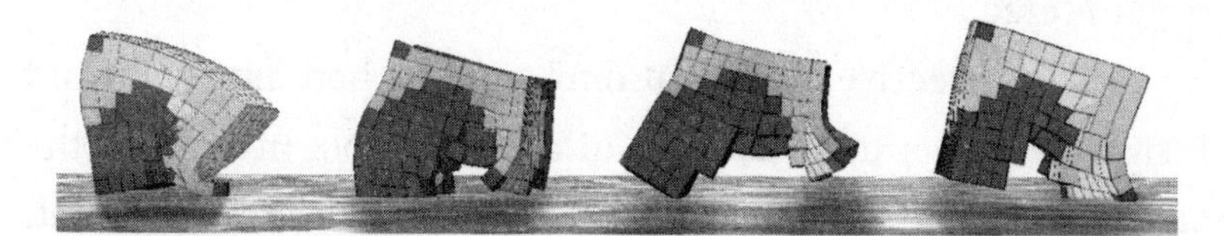

Figure 11.4 The lumbering running style of one example of a "soft robot."

an important role in the immune systems of vertebrates like us. The antibody shown in Figure 11.5 is a protein complex with two outward-pointing "sticky" ends that facilitate the immune response by binding to invaders like bacteria and viruses. The jumble of small appendages seen on the close sticky end should remind you of jumbles you've seen before, as in Figures 3.5 and 10.6. Once again, these are amino-acid appendages. Like all proteins, the entire antibody is bristling with these appendages, most of which aren't shown in the ribbon diagram of Figure 11.5.

You and I owe much to these sticky ends because they have saved our lives, literally, many times. Every time we get an infection, from the common cold to a festering scrape, our bodies go into high gear to fight off microscopic invaders, and antibodies are a key part of winning these fights. Like laser beams guiding missiles to their targets, antibodies *tag* the invaders for destruction, and the high specificity of their sticky ends is what enables them to do this tagging so effectively. To achieve this specificity, our bodies use an extremely elegant version of selective optimization, where billions of variations on the best sticky ends found so far are produced repeatedly, with better sticky ends replacing the previous ones until no further improvement can be made. Adding to the elegance, our bodies retain the best versions of these sticky ends from each of these battles so they can be deployed very quickly the next time the same invader is encountered.

Again, selective optimization is applied here narrowly and insightfully as a tool—as part of a remarkable invention (the adaptive immune system) that cannot have arisen by accident. In every case, this tool proves valuable only by being cleverly employed by someone who knows what it can and cannot do. These masters of selection are the inventors, then. Selection never was. No tool will ever go off and create a world of its own, the way Darwin thought selection did.

Everything fits.

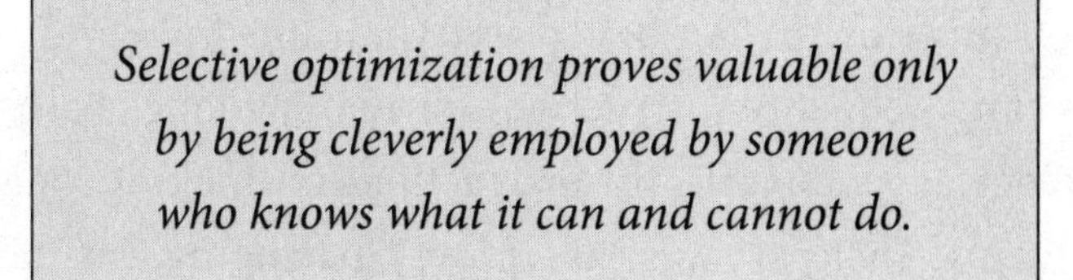

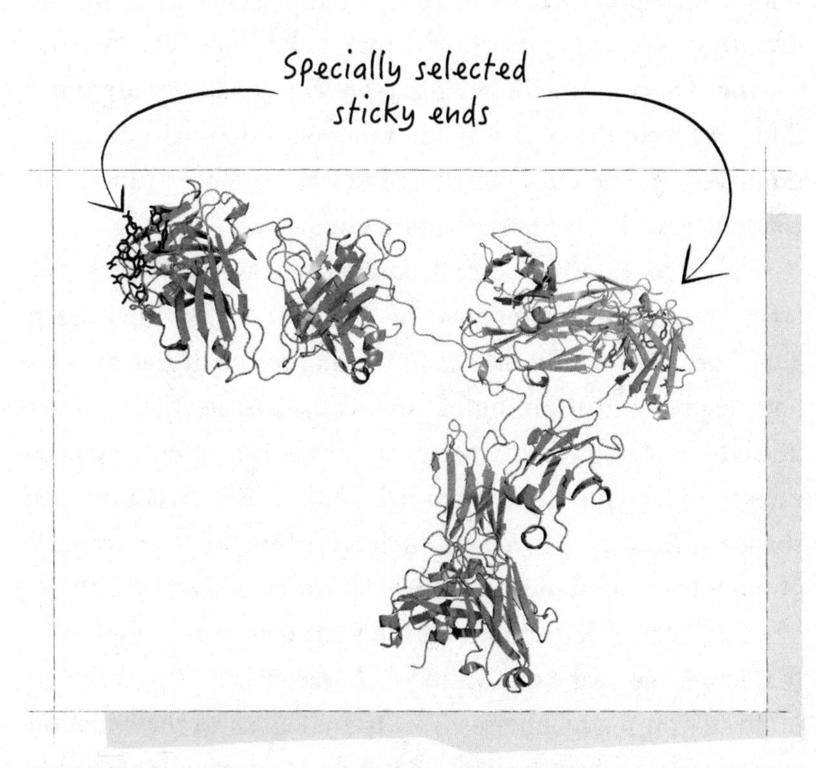

Figure 11.5 The molecular structure of an antibody. Four folded protein chains assemble to form the complete antibody shown here.

Of Language and Life

If you're interested in exploring evolutionary models further, I offer a free computational tool developed at Biologic Institute called *Stylus*. Our objective in developing Stylus was to create a model world that captures important features of the world of natural proteins. In the first place, we wanted a world where genes carry sequence instructions for making long chains, just as biological genes carry the instructions for making long protein chains. That part was easy. More challenging was the goal for these long chains to perform a great variety of actual functions based upon their structures, just as protein chains do. The importance of actual function is that it eliminates all the hand waving that comes with having to *pretend* sequences are functional. Indeed, the only way for a model to do justice to functional coherence is by embracing the whole point of functional coherence, which is high-level function.

After considering several possibilities, we landed on the language analogy represented in Figure 11.6. Like all written languages, written Chinese is functional to the degree that it's both legible and meaningful, and as always, legibility comes down to how well the characters are formed. Written language, then, provides the desired connection between structure (the shapes of lines on paper) and high-level function (the transfer of thoughts from writer to reader). However, whereas the shape of an alphabetic letter does nothing but trigger recognition of that letter, the shape of a Chinese character brings *meaning* to mind (if you read Chinese). This is analogous to the situation with proteins, where each protein molecule performs a distinct function according to the details of its structure.

You recall from chapter 3 that the biological genetic code describes how cells use the sequence information in a gene to attach amino acids in the right order to make a protein. The trick is to read the DNA bases three at a time in codons. We used that same trick for Stylus. Genes in the Stylus world look just like our alphabetic representations of biological genes—

Figure 11.6 Like proteins, Chinese characters must be properly formed in order to work. Keeping in mind that the comparison is meant to be general, not specific, we see rough similarity here between the stroke complexity of a character and the component complexity of a protein. Proteins and Chinese characters both show considerable variation in complexity from one to the next.

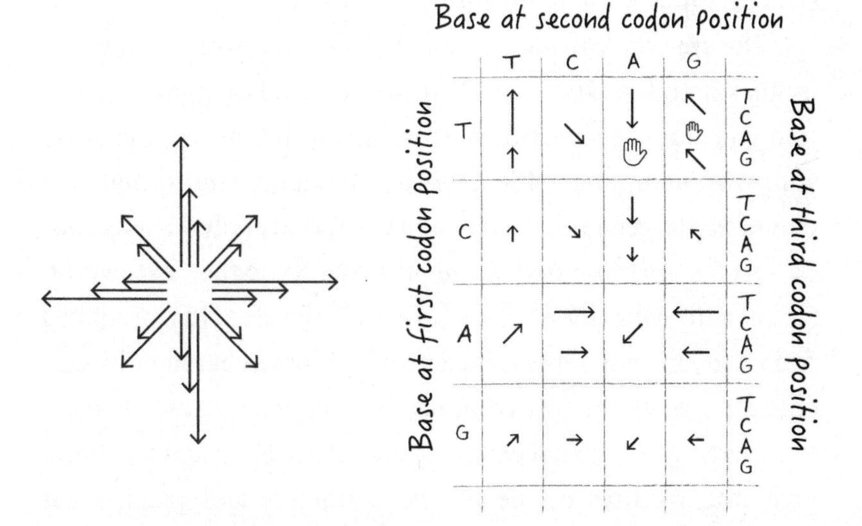

Figure 11.7 The twenty vectors from the Stylus world (*left*) that take the place of the twenty amino acids as building blocks for making long chains, and the genetic code Stylus uses to specify these vectors (*right*). As with the biological genetic code depicted in Figure 3.2, hands indicate codons that terminate a vector chain.

long sequences of the four letters A, C, G, and T—but instead of specifying amino-acid chains, Stylus genes specify *vector* chains. Taking the place of the twenty amino acids are the twenty vectors shown in Figure 11.7. These vectors are joined end to end to form a drawn shape as specified by the encoding gene.

No analogy is perfect, but this one is at least rich enough to be interesting. For example, both proteins and Chinese characters have distinct functional forms numbering in the thousands. And just as a great many amino-acid sequences can construct any one of these protein forms, so too a great many vector sequences can make the same Chinese character. Finally, as we saw in chapter 10, high-level functions like photosynthesis are accomplished when many functional proteins of various kinds are brought together in just the right way. Written language mirrors this beautifully by using its own hierarchical structure to achieve high-level functions.

The free tool, made possible by the hard work of my colleagues Brendan Dixon and Winston Ewert, is a Stylus application you can use to perform experiments at your own pace on your own computer.[12] The application enables you to mutate a parent Stylus gene in a variety of ways and to apply various conditions for selecting one of the offspring to be the next parent. Stylus automatically assesses how well the drawing produced from a given gene represents a given Chinese character by calculating a score ranging from nearly 0 (very poor resemblance) to 1 (perfect resemblance), and it provides rich visual feedback, including pictures of the drawn characters and graphs that show how scores change over the course of an experiment.

As with most tools, Stylus can be used at a range of levels. If you just want a better grasp of how the genetic code works or how mutations affect the instructions carried on a gene, Stylus is a good visual hands-on tool for this. Or you may be much more ambitious. We used Stylus to build a model genome analogous to a small bacterial genome,[13] and between the freely available gene sequences from this genome and the freely available Stylus application, there are plenty of interesting research questions that can be tackled at any level.

For a quick example, recall from chapter 7 the two proteins that had weak penicillin-inactivating ability. One of these proteins was dramatically improved by repeated rounds of mutation and selection, whereas the other wasn't. The difference came down to whether the signal being homed in on was from an actual enzyme—a protein whose structure is specially suited to penicillin degradation—or from something much less special that happened to help the natural breakdown of penicillin along slightly.

Figure 11.8 shows the results of a similar comparative experiment performed with Stylus.[14] Instead of proteins, we started with two vector chains, both acquired by mutating a gene in the Stylus genome that produces a character meaning *section* (段). As in chapter 7, one of these mutant chains was much more severely disrupted than the other, which can be seen by comparing the two drawings in the lower left to the ideal shape. Although the computer gave very low scores to both mutants, we see that the black one was at least partially recognizable, whereas the gray one looked like a random scribble. In fact, the black mutant was structurally complete in the sense it had all nine strokes present; the problem was that some of the strokes had shifted out of their correct locations (the vertical one having shrunk as well). As the black homing curve shows, those imperfections, though substantial in effect, could all be corrected by a series of single mutations, each of which improved the score. The gray scribble could also be adjusted by selecting changes that improved its score, but in this case there was so little connection between the score and actual readability that nothing remotely legible resulted. As with the proteins in chapter 7, selective homing was of value only when functional coherence was already present in large measure.

Nothing evolves unless it already exists.

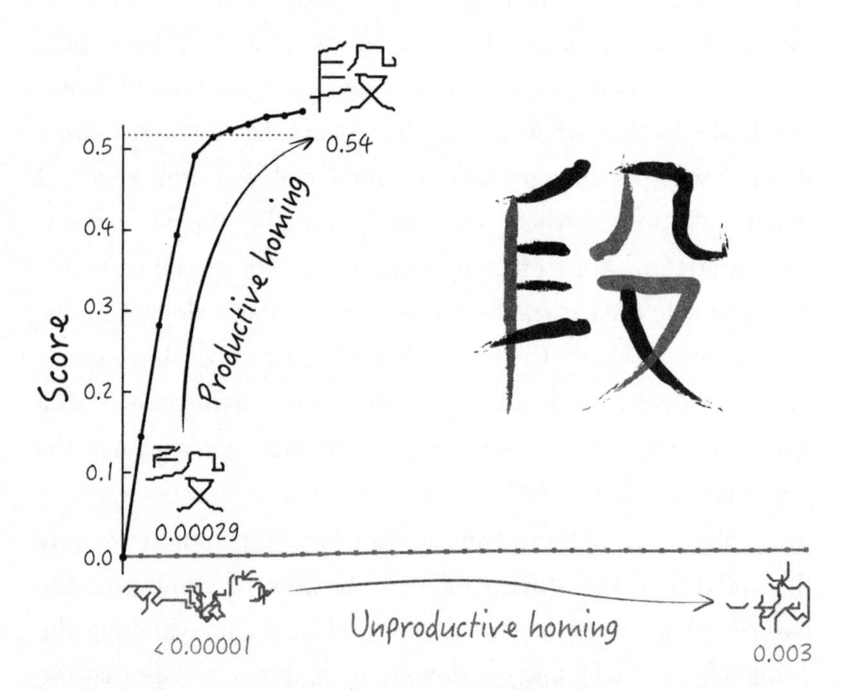

Figure 11.8 Selective homing in the Stylus world. The 段 character consists of nine strokes, as shown with brush strokes. Stylus applies a mathematical test to each vector chain to compute a score, ranging from 0 to 1 (scale on left), which tells how well any vector chain represents the ideal form. Using this mathematical definition of legibility instead of human judgment enables us to take advantage of the much greater "reading" speed of computers. In order for this to represent human perception reasonably well, human assessment of legibility was used to calibrate the scoring. Stylus performs homing experiments automatically by producing and scoring all possible genes that differ from the parent gene by one base (A, C, G, or T) and then selecting the one with the highest score as the new parent. This is repeated until no further improvement is possible. The dotted line shows the score (0.52) of the original gene (designated 6BB5–02), which is included with the Stylus application, along with all the genes from the published genome.

CHAPTER 12

LAST THROES

Having hopefully persuaded you along our journey that our design intuition has triumphed over the evolutionary story, I now want to enlist your help. The truth we've arrived at is important enough that we have a responsibility to stand up for it. Think of this as a movement, not a battle. When a good movement prevails, everyone wins.

Still, movements involve strategy just as battles do, and momentum is a key part of this. One way for a movement to gain momentum is for those who've joined the cause to see opponents of that cause in retreat. Unlike battles, the hope here is that hands will reach out to those in retreat to encourage them to change their allegiance. To that end, this chapter will focus on multiple fronts where the defenders of materialism and Darwinism seem to be in retreat.

The Retreat from Critical Dialogue

Darwin's explanation of life turns out to have been wrong, but in that, it joins a great many other ideas that have had their useful places in the development of scientific understanding. Originally, at least, Darwin's idea was articulated with enough clarity to ensure that it would ultimately prove true or false. Moreover, Darwin clearly identified the crucial point on which this verdict hung. In his words, "If it could be demonstrated that any complex organ existed, which could not possibly have been formed by numerous, successive, slight modifications, my theory would absolutely break down."[1] That is, if any of the inventions we see in the living world can't be acquired one tiny beneficial mutation at a time, Darwin's hopes for natural selection were in vain. His next words—"But I can find out no such case"—revealed where his sympathies lay, which only confirms he was human.

Somewhere within the long succession of his followers—all bright people coaxed into abandoning their design intuition—Darwin's recognition that his idea of gradual invention was vulnerable to refutation was lost. The historical details of the disappearance are too complicated to be pieced together fully, but this is unnecessary anyway. In general terms, the human factors that put an end to vulnerability are very familiar. The idea of natural selection as genius in slow motion became to biologists part of the very definition of life, and with this elevated status came immunity from criticism. To question the most central axiom of modern biology was, and is, to excuse oneself from the company of modern biologists.

The familiar truth is that on matters we care about, we admit the possibility of being wrong only with some reluctance—more as a way of showing ourselves to be reasonable than as a way of encouraging criticism. If and when that concession seems unnecessary, we're inclined to withdraw it. Then, with the passage of time, we become so comfortable with the absence of open criticism that we're indignant when someone unaware of the ground rules violates them. The natural progression, in other words, is from reluctant acceptance of criticism in principle to resentment of criticism in practice.

As human as this progression is, it's distinctly hypocritical when it takes hold in a community that bases itself on reason and open discourse. Faith communities, being explicit in their commitment to doctrine, are being true to their core values when they correct or remove people who oppose what they once agreed to uphold. But for the scientific community to do likewise, based as it is on discovery instead of doctrine, is to violate its core values. The consequences are always ugly. Lacking any special revelation, science boasts intellectual openness as its core virtue. And what a potent virtue this has proved to be! But when openness gives way to dogma on any particular scientific claim, we're left with something more like bad religion than good science.

To spot one of these ugly examples, look for two telltale signs. The first is official denouncement of any idea that poses a threat to the dogma, and the second is a culture of disdain for that threatening idea. On the first point, take a look at the Wikipedia page titled "List of Scientific Bodies Explicitly Rejecting Intelligent Design," where you'll find names of over a dozen academic and scholarly organizations in the United States that have issued statements opposing ID, along with several others outside the United States or of international composition. Among those listed are two of the world's most highly esteemed scientific organizations: the Royal Society of London and the National Academy of Sciences of the United States—impressive opposition, to say the least.

These being scientific organizations, they don't want to appear to have rejected ID on doctrinal grounds, so their denouncements assert that ID can't be given a place at the table of scientific discourse because it's fundamentally religious. Ironically, though, their anti-ID activism has a quasi-religious character of its own. If they actually believed the question of whether life is intelligently designed to be outside the purview of science, they would take no position on the answer. But they *do* take a position. The Royal Society of London has officially declared that "the theory of evolution is supported by the weight of scientific evidence; the theory of intelligent design is not." Likewise, the world's largest scientific society, the American Chemical Society, has urged educational authorities to "affirm evolution as the only scientifically accepted explanation for the origin and diversity of species." And the Royal Astronomical Society of Canada "is unequivocal in its support of contemporary evolutionary theory that has its roots in the seminal work of Charles Darwin and has been refined by findings accumulated over 140 years."[2] Nowhere under that proud Darwinian flag that flies over the modern academy will you find an institutional declaration to this effect: "Although life may well be the work of an intelligent designer, this is not a matter that science can address." That can only mean one thing: the anti-ID activism comes down to a doctrinal stance after all.

That the hoisting of the materialistic Darwinian flag also ushered in a culture of disdain for threatening ideas like ID becomes evident when you look a little deeper than the position statements. Intelligent design is mentioned quite frequently in Darwinian science journals, but always negatively and often with some expression of condescension or contempt. Within these otherwise scholarly pages you'll find intelligent design described—apparently with editorial approval—as a "myth,"[3] as an "attack on biology,"[4] as an "intellectual virus,"[5] as an "insidious movement,"[6] as "the pseudo-scientific face of religious creationism,"[7] as something that "threatens all of science and society,"[8] as "a retreat to the dark ages,"[9] and finally (space not permitting a full catalog of anti-ID epithets) as "*terrifying*"[10]—"like Frankenstein's monster."[11]

Hmm.

Despite the complete rationality of the case for the intelligent design of life, there's just no way to make that conclusion acceptable to people who want to believe science has disproved God. The truth is quite the opposite, which is distinctly uncomfortable for some. Perhaps this explains why some people react so viscerally—not that ID is illegitimate, but just the opposite: it's *painfully* legitimate.

THE RETREAT FROM DARWINISM

As telling as the retreat from critical dialogue is, there are several other equally sure indicators that the search for a natural explanation of living things has come up empty. Perhaps the most striking of these is the repeated acknowledgment from scientists closest to the subject that Darwin didn't actually achieve such an explanation. This is the gaping hole we encountered in chapter 7. In his 1904 book *Species and Varieties: Their Origin by Mutation,* the great Dutch botanist Hugo De Vries stated the deficiency as follows:

> In indicating the particular means by which the change of species has been brought about, [Darwin] has not succeeded in securing universal acceptation. Quite on the contrary, objections have been raised from the very outset, and with such force as to compel Darwin himself to change his views in his later writings. This however, was of no avail, and objections and criticisms have since steadily accumulated.[12]

So the sudden favorable turn of scientific opinion Darwin described in the sixth edition of his book, where biologists went from believing in the separate creation of each species to accepting his "great principle of evolution," wasn't accompanied by general acceptance of natural selection as the cause.[13] Accordingly, De Vries ended his book with a memorable quote describing what we are calling the gaping hole: "Natural selection may explain the *survival* of the fittest, but it cannot explain the *arrival* of the fittest."[14]

Despite huge boosts in subsequent decades from the development of a mathematical theory of natural selection and the discovery of DNA as the genetic material, Santa Fe Institute scientist Walter Fontana and Yale biologist Leo Buss acknowledged in 1994 that the hole in evolutionary theory was still unfilled. Repeating De Vries's memorable quote ninety years later, Fontana and Buss titled their paper "'The Arrival of the

Fittest': Toward a Theory of Biological Organization." It begins with a big concession:

> The formal structure of evolutionary theory is based upon the dynamics of alleles [i.e., gene variants], individuals and populations. As such, the theory must *assume* the prior existence of these entities.[15]

Don't miss the significance of this. Because all living things are among "these entities," Fontana and Buss are admitting here that modern evolutionary theory doesn't actually explain the origin of new species or even the origin of new genes. Instead, "present theory tacitly *assumes* the prior existence of the entities whose features it is meant to explain."[16]

If you're wondering why some scientists get away with such startling frankness while others are censured or excommunicated, it all comes down to whether the critic is seen as a friend of the greater cause. Scientists can say what they want about the state of evolutionary theory if their allegiance to scientistic materialism is intact, and the best way for them to demonstrate that is to claim to have *filled* the hole, or at least made decisive progress in that direction. As with road repair, you're allowed to use the jackhammer with impunity as long as you patch everything up before you leave. Using that strategy, Fontana and Buss offered their deep criticism as a way of introducing a theory that, by their account, explains how "self-maintaining organizations arise as a generic consequence of two features of chemistry, without appeal to natural selection." In other words, Darwin was wrong, but life is still the expected outcome of blind chemistry, so all is well under the materialist flag.

Like those two scientists, a great many others have tried to patch the hole in Darwin's theory over the years, but none of these patches have proved very durable. Interestingly, progress in biology seems to make the situation worse. The genomic era, a period of unprecedented progress, was just getting under way when the paper by Fontana and Buss appeared. Twenty years later came a book by Swiss evolutionary biologist Andreas Wagner. So if the patch offered by Fontana and Buss were sound, Wagner would have been in a good position to affirm this. Instead, he reaffirmed the existence of the gaping hole, as evident in his title: *Arrival of the Fittest—Solving Evolution's Greatest Puzzle.* Echoing his predecessors, Wagner concedes that "natural selection can *preserve* innovations, but it cannot *create* them." After this, he says:

> To appreciate the magnitude of this problem, consider that every one of the differences between humans and the first life forms on earth was once an innovation: an adaptive solution to some unique challenge faced by a living being.[17]

What Wagner calls *innovations* I've called *inventions,* but the point is the same and it applies as much to spiders and whales and orchids as it does to humans. Of the countless remarkable inventions on display in the countless remarkable forms of life, natural selection explains *none* of them.

Wagner gets away with this devastating critique of Darwinism the same way Fontana and Buss did—by offering his idea of a solution. Certainly, if Wagner's solution really filled the hole, it should be heralded as a remarkable achievement, in that it brought 155 years of failure to an end. Being familiar with the subjects he deals with, I could tell you why I think he didn't succeed, but in effect I would be asking you to trust me over him, which none of us should find satisfactory. Instead, my whole purpose has been to equip you to trust your own design intuition.

Wagner ends his book with this summation of his thesis:

> With a limited number of building blocks connected in a limited number of ways, you can create an entire world. Out of such building blocks and standard links between them, nature has created a world of proteins, regulation circuits, and metabolisms that sustains life, that has brought forth simple viruses and complex humans, and ultimately, our culture and technology, from the Iliad to the iPad.[18]

His first sentence—where the actor is *you*—is surely true. Wagner wrote his book the same way I'm writing mine—by connecting the twenty-six alphabetic building blocks. Software developers connect commands and then compile these to get long lists of connected zeros and ones. The periodic table of elements organizes the fundamental material building blocks in a way that explains the standard chemical links between them. Those elements and their connections make possible absolutely every physical thing we humans make, including the iPad. But as necessary as all these building blocks are for our inventive activities, *they* are not the inventors. *We* are.[19]

Consequently, Wagner's second sentence—where the actor is *nature*—sounds like a fairy tale to everyone whose design intuition is still intact. And again, unless we've gone very wrong in our thinking, it *should* sound like a fairy tale. Alphabet soup is chock-full of building blocks, but nature is so clearly incapable of doing what we do with building blocks that we knew immediately the account of oracle soup couldn't be true. As we realized back in chapter 2, an irrefutable demonstration of that mysterious soup would only convince us that an invisible *someone* is arranging the letters.

The mere *tale* of oracle soup, however, sends no shivers down our spines for the simple reason that we have no reason to regard it as true. Likewise for all these tales of nature inventing. After more than a century and a half of these stories, surely we're entitled to ask for something more. Talk has its proper place in science, but to those hoping to convince everyone that accidental invention is possible, I say—If you merely *tell* us that plants happen when light shines on random atoms or that nature created a world of proteins, the response will probably continue to be disappointing. *Show* us such magical things, and you will have our rapt attention. Give us a demonstration that passes the hat test with flying colors. We will still be puzzled by your insistence that magic should be regarded as ordinary, but you *will* have our attention.

The Retreat from Testability

Because the arguments and evidence run counter to Darwin's idea, perhaps we shouldn't be surprised to see defenders of that idea shrinking back from scientific discourse. Figure 12.1 illustrates their predicament. To attribute the invention of all complex life to a natural mechanism set in motion with the earliest simple life is to ascribe astonishing creative power to that mechanism. Yet when my colleagues and I challenge this evolutionary mechanism to invent on a far less impressive scale—by altering an existing enzyme to perform a new function—we find that it fails. It's hard to exaggerate the magnitude of this contradiction. Imagine a group of people insisting that a certain

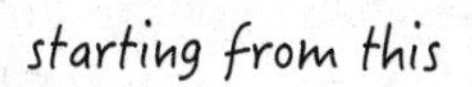

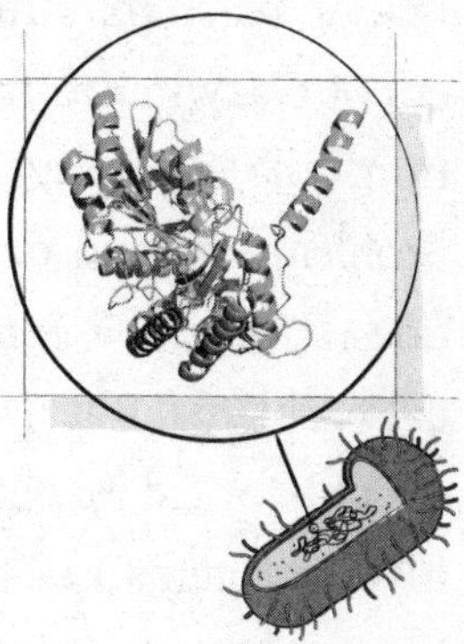

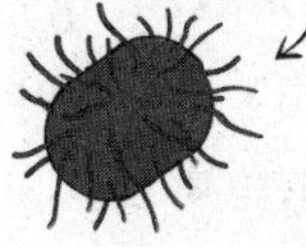

Figure 12.1 The striking contrast between a very minor invention that isn't feasible by Darwinian evolution (*top*) and the profound inventions that Darwin assumed were feasible (*bottom*). Shown at the upper left is enzyme *B* from chapter 6. In a bacterial population that already makes the very similar enzyme *A* (*upper right*), evolutionary conversion of *A* to *B* appears not to be feasible based on our studies. The lower half of the figure illustrates invention at the much higher level of complex life forms. Here, evolution would have had to invent *all* life from some early bacterial species supposed to be the ancestor.

man can jump to the moon. We, being skeptical, challenge this man to dunk a basketball, and we find that he comes well short of reaching the rim. When we publish our finding, we get lots of complaints, all of the kind "We never said he could dunk a basketball . . . or at least not *that* basketball, on *that* rim!"

In other words, recognizing the difficulty of getting their man to reach the rim, to say nothing of the moon, most defenders of Darwin are trying very hard to embrace the contradiction. The thinking is that evolutionary ineptness in solving problems that no one claims were solved in the history of life doesn't preclude competence in solving the supposed historical problems. But the matter of competence must take priority over historical assumptions. That is, the claim that evolution *did* invent proteins, cell types, organs, and life forms is scientifically legitimate only if we know evolution *can* invent these things. Consequently our demonstration of evolutionary incompetence for an example of the least of these inventions—a new function for an existing enzyme—undercuts the whole project of inferring evolutionary histories. If nothing *can* evolve its way into existence, then nothing *did*.

> *The claim that evolution* did *invent proteins, cell types, organs, and life forms is scientifically legitimate only if we know evolution* can *invent these things.*

When that previous statement is no longer presumed true, we know we've reached the final stop in a staged retreat from testability. As I mentioned near the end of chapter 6, this is where the debate now stands. The current stance is that evolution was so successful that it perfected life to the point where modern forms no longer evolve, making the whole process even further removed from the category of observable phenomena. By this view, direct observation of evolutionary invention would require access to a world that no longer exists, and because evolvability is presumed to have been a characteristic of that world, any attempt to reconstruct a piece of it in the lab would be judged by whether evolvability was confirmed. In this way refutation seems to be forestalled, but not without considerable cost. To put it bluntly, evolutionary theory has become immune to refutation in much the same way that the stump of a tree has become immune to further pruning.

Strangely, after all the "anti-science" insults that have been directed at proponents of intelligent design, we seem to be among the few who are interested in using science to settle the matter.

The Retreat from This Universe

In 2007, Eugene Koonin, a prominent evolutionary biologist at the National Center for Biotechnology Information in Maryland, gave scientists in his field a double shock. The first shock was his frank concession that the origin of the first cell carrying genetic instructions for making proteins is "a puzzle that defeats conventional evolutionary thinking."[20] Having deployed this jackhammer, he was obligated to fill the resulting hole, which he attempted to do in a most unconventional way. Koonin delivered his second shock by appealing to *cosmology*—

the study of the origin and behavior of the universe as a whole—to patch everything up. Specifically, to dispense with the problem of fantastic improbability, he leveraged the idea of an infinite *multiverse,* which we may think of as an infinite set of actual universes, ours being one.

Understandably, most people consider this supposed multiverse to be so far removed from real experience that they have a hard time taking it seriously. But while that skeptical stance should inform the discussion of where the boundaries of science lie, *truth* is a bigger and more profound subject than science. For my part, although I reject the existence of other universes, I'm not content to do so simply on the grounds that we can't verify their existence, because it seems equally true that we can't verify their nonexistence. The better question is whether the hypothetical possibility of an infinite multiverse should change the way we explain life in *this* universe.

Koonin's reason for thinking it should is based on a concept called the *anthropic principle.* Autobiographies of the "I lived to tell the tale" kind show how this principle works. In these books, the author recounts circumstances where death seems virtually certain, and yet the very fact that he or she lived to tell the tale assures us that the odds of surviving, however slim, were somehow overcome. In the most extreme cases, many of us would say that God intervened and therefore chance had nothing to do with the turn of events, but the point is that no matter how impossible the situation seems as we read the account, we know that some remarkable occurrence must have averted death—otherwise the book wouldn't have been written.

The anthropic principle applies a similar idea to our existence. If we suppose, even hypothetically, that our universe is just one of infinitely many parallel universes, and that conditions vary from one universe to the next, such that all the innumerable physical possibilities become actual *somewhere* out there in the multiverse, then might we humans be like authors who lived to tell the tale? To follow the reasoning here, start by supposing that the probability of a universe producing intelligent beings like us by accident is greater than zero (we'll return to this later). Using *gazillion* as a placeholder for some very large number, we'll say this probability amounts to one in a gazillion. It follows, then, that for every gazillion universes, *one* is expected to have intelligent beings who came about by accident. And because an infinite collection of universes has an infinite number of gazillions within it, it follows that there should be an *infinite* number of these very special one-in-a-gazillion universes that are home to thinkers like us—not by the hand of God but by the raw power of infinitude.

But what seems at first glance to be at least a provisional theoretical possibility doesn't square with reality. To see this, ask yourself what we should see in our universe if things really were as we have supposed. The answer is that we, as beings who wonder about our origin, should see the most bare-bones circumstances for wondering to be possible. The rationale for this answer isn't complicated, but the whole scenario is so strange it may take some effort to get your head around it. As with the search spaces we discussed in chapter 8, this hypothetical multiverse would consist almost entirely of uninteresting alternatives. These would be normal universes where the fantastic improbability of accidental invention equates to physical impossibility, so no invention occurs. But if we assume, first, that it's not categorically impossible for physical processes to produce beings capable of wonder and, second, that the infinite multiverse is real, then our present act of wondering could be explained by our universe being one of those fantastically rare universes where the staggering improbability of wonderers being invented by accident just happened to be overcome.[21] Like the autobiography, our existence would be the proof.

Now, here's where evolution comes in. *If* it were true that evolution works as a brilliant inventor and that intelligent beings like us are among the things it can invent, then I would agree with Koonin. The most bare-bones explanation for us would be that simple cellular life was formed against all odds, and evolution took over from there. So Koonin's appeal to a multiverse as a way of explaining how the fantastic improbability of that first cell was overcome is perfectly consistent with his set of assumptions.

However, once we realize how *incompetent* evolution is as an inventor, this whole multiverse explanation collapses. We do indeed find ourselves in a world where the individuals of one species—ours—wonder how everything came to be, but a big part of our wonder has to do with the obvious fact this is anything but a bare-bones world. Quite the opposite. So since every one of the biological inventions that surround us is fantastically improbable, with evolution explaining none and the multiverse hypothesis explaining only those absolutely necessary for wondering to be possible, we conclude that this hypothesis fails to explain what we see. Conceivably, we could have found ourselves wondering on an austere planet populated by little more than lonely thinkers whose bodies are capable only of those functions absolutely necessary for thought. And because that kind of planet is *vastly* less improbable than this sumptuously appointed five-star accommodation we call Earth, we certainly *would* have found ourselves there rather than here if we really were accidents of nature.

That we are here instead assures us we are not.

THE ELEPHANT IN THE ROOM

Because reality can't ultimately be grounded in physical things, materialism always fails when we ask big questions of it. This categorical inadequacy of the physical realm makes the number of physical universes irrelevant. Physical processes simply can't be the basis of everything, no matter how much room we give these processes to work.

A similar principle holds for our *understanding* of reality. Contrary to the claim of scientism, we can't ultimately base our knowledge of truth in science. To see this, let's momentarily adopt the mind-set of an absolute skeptic—someone who doubts everything that can be doubted. No one really is an absolute skeptic, and most of us never go to the trouble of even contemplating absolute skepticism. It will be worth doing for a moment, though, just to see how hopeless it would be to make skepticism our top priority.

Think with me for a minute in the first person. How do I know I existed one minute ago? Is it enough for me to say I remember the past and see evidence of my past? Usually this is enough, of course. At the end of a workday, I always find a familiar car in the spot where I recall parking my car, and this seems to confirm my recollection. While operating in absolute-skeptic mode, however, I have to admit that these ties to my past are nothing more than present impressions, and I can't convince myself that my present impressions are infallible. I believe them, but I also find myself revising these beliefs quite regularly, as when waking from a dream, for example. So how do I know this whole life experience isn't a dream that just popped into existence a moment ago? Again, I find that I'm content to believe otherwise, which is a very good thing considering I can do no better. Neither can you. We *must* take some essential things on faith because there really is no alternative.

My point is not that anyone is or ought to be an absolute skeptic. Rather, my point is that *faith alone* is what rescues us from the futility of absolute skepticism. If you think science can come to the rescue instead, ask yourself what confidence you can place in science without presupposing that you *did* exist a minute ago. If your entire past is an illusion, how can this thing you call "science" not also be an illusion? The truth is that science can't even *conceivably* give us anything more certain than the faith we place in the essential propositions undergirding science, which means science will never be the primary path to knowing, much less the only path to knowing. Faith has always been more fundamental to human knowledge than science, and this will never change.

What makes science so compelling is that we all do accept the essential propositions. And when we add nothing to those essentials, common science and common sense naturally lead us to attribute life to God, as even the children of atheists do. I can vaguely imagine a version of reality where God exists but science and reason are silent about his existence. The conceivability of that imaginary world makes it all the more striking that this world is so different.

Here the silence is broken.

Science can't even conceivably *give us anything more certain than the faith we place in the essential propositions undergirding science, which means science will never be the primary path to knowing, much less the only path to knowing. Faith has always been more fundamental to human knowledge than science, and this will never change.*

WHERE IT ALL LANDS

There's no way around the fact that everything resembling earthly life requires high-level functional coherence. Nor is there any way around the fact that this makes the sum total of all possible things that would be recognizable as earthly life impossibly rare. In the language of chapter 8, these possible life forms are hopelessly lost in the nearly infinite space representing the ways matter could be mindlessly arranged. What the inventor can do—seeing possibilities that are otherwise not there and seizing opportunities that only exist because they are imagined—cannot be done by accident.

Yes, Darwin's idea falls when we grasp this, but so does every attempt to pretend that life just happened, no matter how sophisticated it may sound. If we try to avoid God by supposing all the necessary elements for each evolutionary step just happened to be at the right place at the right time, against all odds, then we only push his creative work back from the creatures themselves to the circumstances that brought them about. If we try to avoid God by supposing life came to Earth from outer space, we only push his actions to another planet or another galaxy. If we try to avoid him by supposing life unfolded from the initial conditions of the universe or from the laws of physics, we make these things so ingeniously life-directed that we have only pushed his action back in time. And if we try to avoid him by imagining our universe is only one among an infinite number of universes, he shows himself to be present here nonetheless. That he has acted is plain to see, and no theory can erase what we see.

All of this follows from the fantastic improbability of life having happened by accidents of any kind. All of it declares God's presence and involvement in our world—breaking the silence, shouting to anyone who has ears to hear.

And as if that were not witness enough, there is more.

CHAPTER 13

FIRST WORLD

I haven't shied away from naming God as the knower who made us. I see no other way to make sense of everything we've encountered on our journey. By recognizing that life can't be an accident, we're affirming that life was intended, and considering the stunning genius with which this intent was carried out, I'm compelled to see God as the genius behind it.

Still, thinking in particular of Thomas Nagel, I want to put my finger on what it is that makes the alternative hypothesis of a purposeful but *impersonal* natural order untenable to me. As we approach the end of our journey, the time has come to address this.

Nagel is drawn to the subject of mind, and so am I. Perhaps the most awe-inspiring aspect of the outside world is that we each view it from our own *inside* world. Not only that, but we are active participants in the outside world. Like operators of diesel-powered earthmovers, we each have a seat on the inside from which we see the world outside while also acting to change it. How is this possible? Clearly it is, or we wouldn't be here to talk about it. Still, once we get past its familiarity, this marvelous truth ought to stir wonder and amazement in us.

The children whose simple view of life has proved superior to the view endorsed by the Royal Society and the National Academy also have a simple view of consciousness. Their view begins to take shape in infancy with games like peekaboo, where small hands over small eyes form a screen that momentarily isolates the inside world from the outside world. With the screen in place, the view from inside is one of darkness and mounting anticipation. Then the hands are flung from the eyes and the anticipation is rewarded, always with shrieks of approval. Through countless learning moments like this, children build a connection between their inside world and the outside world, a connection far more profound than anything technology has given us.

Further exploration deepens the connection. The child begins to recognize that certain special participants in the outside world (Mommy, Daddy, sister, brother) are also viewing it from an inside world—their *own* inside world. This understanding is imperfect at first. The child initially thinks that by covering his eyes he closes off the outside world for everyone. Later he learns that Mommy's eyes are the window to her inside world just as his eyes are the window to his, and so the child's internal model of reality is refined.

With increased internal understanding comes increased external expectations, along with consequences for successes and failures in meeting those expectations. Learning from both, the child eventually attains full self-awareness, making decisions with the understanding that they have an effect on the outside world, that those effects will be observed by other people, and that the observers may respond with decisions of their own. Of course, in coming to this view of reality no one feels as though they're doing science or philosophy. Indeed, this commonsense view seems so natural that few of us give it a second thought.

THINKERS, THOUGHTS, AND THINGS

When we ponder this view for a moment, it seems to place the components of reality into three categories. I'll call them *thinkers, thoughts, and things*. The outside world consists entirely of things (galaxies, atoms, trees, computers, etc.), whereas each inside world consists of the mental space in which one thinker has thoughts. I referred to this personal space as a workshop at the close of chapter 10 because each person works upon his or her own thought projects within it.

This view stirs controversy when we take it back to the question that started our journey: *What is the source from which everything else came?* Immediately another interesting question arises. If reality presently consists of thinkers, thoughts, and things, which of these three should we see as being *primary* in the sense of having been the source of the other two? As I mentioned at the outset, materialists hold *things* to be primary, whereas theists hold *thinkers* to be primary—most notably the divine thinker we call God. Hence the tension.

The struggle for materialists has always been to explain how *things* could really be primary. For things even to be the source of all *things* seems impossible. The spider's spinning apparatus is certainly a thing, but because it's one of those special things we call inventions, we've concluded that it can't have originated by accident. Only a *thinker* could have brought spinnerets into existence, and a highly clever thinker at that.

Up to this point, our discussion has focused almost entirely on that one key failure of materialism: its inability to explain inventions. Nagel and a good many others have pointed to another failure, which happens to be the one my own reflections first turned to during my undergraduate days. More profound than the failure to explain invention, this other failure is that materialism is *categorically* incapable of explaining either thinkers or their thoughts. As a student of the physical sciences, I realized that however powerful physical and chemical descriptions of matter are within their own domains, they can't possibly describe the most important aspects of *us*. One of my handwritten notes on that student bulletin board mentioned back in chapter 4 summarized my reasoning as follows:

> Physical systems are governed by physical laws. With our minds we are able to control our physical bodies. Our minds can take precedence over physical laws and are therefore nonphysical. That which is constrained by physical laws cannot give rise to something that takes precedence over those laws. Therefore, man did not evolve from the physical.

In other words, the problem as I saw it was not merely that the mind is currently beyond physical description but rather that the mind is categorically *above* physics. The properties of matter make all mere *things* behave the way they do, but somehow we stand above that. We are not mere things. Within the limits of our capabilities, we do whatever we want to do without answering to any equation.

But while we are masters over the physical in that sense, we are slaves in another. Our bodies are physical things, subject to physical needs and vulnerable to physical conditions. Without food, water, and rest, they cease functioning properly, and our minds are quick to follow. In fact, our minds are particularly sensitive to certain material influences. The most active thinker among us can't stand up to a dose of propofol, a common drug for inducing general anesthesia. So the same minds that spend their waking hours manipulating matter are rendered completely inactive by a small amount of matter of a certain kind. The point is not that we stand wholly above the material world, as God is said to, but rather that we occupy a position that so categorically defies material explanation as to refute the materialist position, *over and above* the refutation that most of this book was devoted to.

A simple thought experiment should convince everyone of this. Imagine yourself seated inside a brain-imaging laboratory surrounded by complicated equipment, some of which is connected to you by means of wired probes stuck to your scalp and forehead. You are fully conscious, not alarmed in the least by your surroundings, in no need of sedation (we can do this in a thought experiment). In fact, you are calmly conversing with the brain scientist standing before you in his white lab coat. You were thoroughly intimidated by him when you were escorted into the lab, but the conversation soon took such an amusing turn that all intimidation vanished.

> "I'm still trying to pinpoint exactly what you mean when you say *two*," he says with more than a hint of frustration.
>
> "I keep *telling* you what I mean. *Two* is the next whole number after *one*—one more than one."

> "Which is one more than none."
>
> "Exactly."
>
> "Yes, well, that all sounds very nice, but I can assure you that every one of your thoughts is nothing but a physical manifestation of this mass of neurons that sits inside your skull, and I can likewise assure you that I am recording and imaging absolutely *everything* going on in there—*very* expensive equipment this is, the very latest—and yet whenever I show you something on this display that looks very promising to *meeee . . . youuuu* keep insisting it is not at all what you mean by *twoooo*. I might have overlooked this if we had fared any better with the other words we tried: *circle, triangle, line, around, between, love, hate, true, false, one,* and *none*. But as things stand, I am beginning to think this whole exercise has been a colossal waste of time—*my* time, that is. And no, I most certainly do *not* want to know what *you* mean by *time*!"

Do you see the problem? The meaning we attach to these words is nowhere to be found in a person's brain, or in any other physical location, for that matter. Of course, some of these words may describe various aspects of the brain, but to describe a thing is not at all the same as to be *indistinguishable* from it—*identical to it. Yes, two* happens to be the number of cerebral hemispheres, but it is also the number of Martian moons, the number of sides to a coin, the number of Nobel Prizes awarded to Fred Sanger, and the number of sequels to *A Fistful of Dollars*. In the larger scheme, all of these are small, contingent realities, whereas *two itself* towers above them as a permanent, universal, necessary reality.

Or at least I can make no sense of anything apart from this view. To say that a statement is *true* is to say something significant precisely because *true* is another of these necessary realities. If *true* were anything less—a physical process in a brain lobe, or a substance that could be packed into 50-milligram tablets, or an object that could be photographed—then the lofty meaning we attach to truth would be a mirage, and this activity we're doing right now called *reasoning* would instantly collapse into a heap of pretentious nonsense. We vigorously insist that certain claims are true and others false because we believe that the distinction goes much deeper than whim. But if truth were nothing more than, say, a particular pattern of neuronal firing in one's frontal lobe, then insisting we prefer truth over falsehood would be like insisting we install the toilet paper roll this way and not that way. Much ado about nothing.

Indeed, our very sanity is at stake here. If the meanings we ascribe to nearly every word we use are as completely mistaken as the scientist in our thought experiment suggested, then every sentence uttered or written throughout the history of human thought has been fundamentally confused, and in thinking otherwise we show our *thoughts* to be every bit as confused as our words. In fact we couldn't even *have* thoughts. The conceptual realm is, after all, where we *think* we do our thinking, so if this realm doesn't really exist in its rightful position above the physical realm, then we're badly deluded.

But once we reject the materialistic premise, we see irony instead of insanity. The reality of truth makes reason possible, which in combination with physical observation makes science possible. No one should deny the importance of science, but neither should anyone deny the importance of the more fundamental realities that lend meaning to science. By doing just that, materialism and scientism invalidate the very discipline they seek to elevate.

In the end, it seems the children are right again. The inside world is every bit as real as the outside one. Consciousness and free will are not illusions but foundational aspects of reality, categorically distinct from the stuff of the outside world. Following the children, if we allow ourselves to see the outside world as an expression of God's creative thought, everything begins to make sense. Far from peculiarities to be explained away, consciousness and free will are at the very center of all reality, just as they are at the center of us. We love to think and create and express ourselves because we were created to do so by a God who has surrounded us with exquisite proof that he loves these same activities.

The Primacy of Personhood

Can there be any other coherent view of reality, then? On this point I find Thomas Nagel's transparency extremely refreshing. As I mentioned in chapter 1, he wants there to be an alternative—a fully coherent view of reality where God is absent. I have never felt that desire, but I can relate to the underlying "cosmic authority problem" Nagel describes, at least to a degree. Hasn't each of us wanted to be his or her own authority, accountable to no one but ourselves? I've never wished God out of existence during those moments, but I certainly have felt his presence to be something of an inconvenience, to put it mildly. All of this I regard as part of the human condition, and being human, I'm wholly sympathetic.

Nevertheless, while I admire the way Nagel thinks and agree with much of his reasoning, I find his atheistic position fundamentally untenable. Nagel affirms that "our clearest moral and logical reasonings are objectively valid,"[1] and he rejects the materialist view on the grounds that it lacks even the categories for treating these reasonings as real. Having dispensed with the dominant academic view in that way, he seeks "alternatives that make mind, meaning, and value as fundamental as matter and space-time in an account of what there is."[2] I would say *even more* fundamental, but I otherwise agree with all of this, as with Nagel's more complete description of what we should be looking for:

> The hope is not to discover a foundation that makes our knowledge unassailably secure but to find a way of understanding ourselves that is not radically self-undermining, and that does not require us to deny the obvious. The aim would be to offer a plausible picture of how we fit into the world.[3]

Well put. And yet it seems to me that Nagel's careful probing of the deficiencies of materialism shows the missing category to be not just moral or logical reasoning but something considerably bigger.

As I mentioned at the start of chapter 10, Nagel is seeking a version of the natural world that produces the things of personhood—consciousness, reason, and moral sense—as part of the expected course of events. In other words, he wants a version of nature that, once accepted, *demystifies* the appearance of all these remarkable inventions we've been considering along our journey, including the higher faculties of humans. Everyone sees the difficulty of constructing what Nagel is calling for, including Nagel himself, but we don't want to dismiss the project just because it looks hard.

I find myself having to reject it for two other reasons. First, when I try to imagine a way of understanding the world that meets Nagel's conditions, I'm left to think that the mystery he wants to eliminate can only be *displaced*. Humans might be the expected outcome if this hypothesized picture of nature were correct, but that only makes the picture *itself* mysterious. Of all the attributes this impersonal thing we're calling nature might have had, *why* would it have had the astonishing attributes needed to produce such astonishing things?

My second reason is that I honestly think Nagel is rejecting the obvious. What is conspicuously neglected in materialism *and* in Nagel's hoped-for alternative to materialism is *personhood*—not as something derived but as something *fundamental.* Indeed, if the aim is to understand how we fit into the world, then the subject of *we* deserves as much attention as the subject of *world.* Materialism has committed the glaring error of adding *we* in only at the very end, as an afterthought, and Nagel should be thanked for pointing this out with admirable clarity. But even he entertains a rather truncated and fragmented notion of *we*, I think.

We know from what we've discussed that the *cause* we owe our existence to cannot be accidental. Nagel agrees, but to avoid a personal understanding of purpose, he seeks a natural one instead, where "things happen because they are on a path that leads toward certain outcomes—notably, the existence of living, and ultimately conscious, organisms."[4] So this *cause* brought us into existence *as if* we were intended, and to do so this *cause* must have been in possession of what amounts to astonishing insight. All this is clear. Add to it the fact that this *cause* must have encompassed the very categories that Nagel addresses—conscious mind and rational faculties and moral sense—and if *I* were in Nagel's shoes, *I* would find the resulting profile to be unnervingly personal.

Indeed, how could something that lacks personhood know the *path* to personhood? How can anything intend to produce persons without first understanding what this *means*?

If the obvious solution to all this is to acknowledge the reality of a personal God, why go to such strained lengths to withhold this acknowledgment? Having now dipped our toes in these intriguing waters of reconciling our inner and outer worlds, *why not just dive in?* In 128 pages Nagel unraveled 150 years of Western thought just by taking seriously a few obvious aspects of humanity. Why not go further? Yes, we are conscious thinkers with a moral sense, but that's hardly an adequate description. We are also friends and lovers and givers and takers and dreamers and visionaries and storytellers and philosophers and advocates and pleaders and sympathizers and sacrificers and accusers and forgivers and warriors and peacemakers and singers and sculptors and painters and musicians and poets and worshippers and dancers and actors and comedians and chefs and winemakers and revelers and rescuers and healers and planners and competitors and risk takers and thrill seekers and explorers and builders and creators and leaders and followers and yearners and regretters and reminiscers and laughers and criers . . . and much more besides.

In other words, the overwhelming richness displayed to us by the outside world is complemented by an equally rich inner experience—almost as if the two were made to go together.

Our cup overflows.

Diving In

Let's take a moment to wade in the shallow end before we dive into the deep end. Have a look at Figure 13.1, where you'll see things that are instantly recognized as manufactured objects, even if we have no idea what they are. They are indeed manufactured—out of a hard silica-based material similar to opal. I'll also tell you that each of them is tiny enough to fit on the end of a strand of human hair. With that I think you'd agree that they are *remarkable* manufactured objects, exhibiting both technological sophistication and elegance of form (Figure 13.2 shows some close-up details). Where did these remarkable manufactured things come from? Are you perhaps picturing a giant nanofabrication facility in Silicon Valley? If so, you should be blown away to hear that these are actually outer casings from single-celled algae called *diatoms*. It's true. The factories that produced these exquisite pieces of tech art *were the individual algal cells that lived within them!*

Sarah Spaulding, an ecologist with the U.S. Geological Survey who has identified dozens of diatom species, describes her love of them as an obsession that began with her first glimpse of them through a microscope. In her words, "I think that if only other people could see diatoms, they would be just as obsessed as me."[5] I know the feeling. Just viewing pictures like these triggered a similar passion in me, almost as though I'm seeing things from another world—the exquisite craftsmanship of an alien superintelligence. How much more exhilarating it must be to know you're the first human seeing and describing one of these extraordinary life forms!

Here's an interesting thought, then: What if we were *meant* to be exhilarated by experiences of this kind? And furthermore, what if we were meant to *know* this? What if this thrill—our inner world becoming ecstatic about our outer world—was meant to be something even deeper and more personal than mere discovery? What if science was meant to

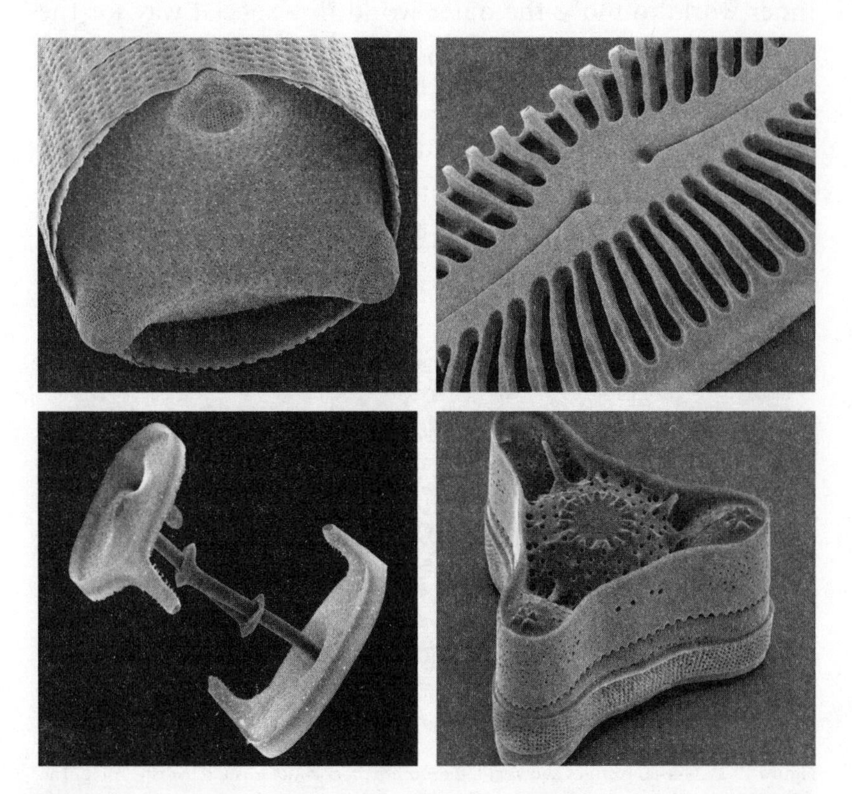

Figure 13.1 A small sample of outer casings, called *frustules,* which exist in about 100,000 distinct forms.

be like geocaching on steroids? What makes finding a well-conceived geocache so delightful is not just the sense of having found something that was hard to find—though that's part of it—but the sense of having found something that was both *meant* to be found and cleverly *made* hard to find. The beach-comber with the metal detector outdoes the geocacher in terms of monetary gain, but the geocacher comes away with something much more valuable than lost change and jewelry. The geocacher comes away with a personal connection to others who are unseen but tangibly sensed and appreciated—another of those beautiful moments when one person uses his or her inner world to move the outer world in a special way for the express purpose of touching another.

Shrieks of approval.

You and I may never have an opportunity to collect diatoms from remote waters or to place them under the beam of a powerful electron microscope. That's okay. Seize rare opportunities like those as they arise, but be assured that we are surrounded

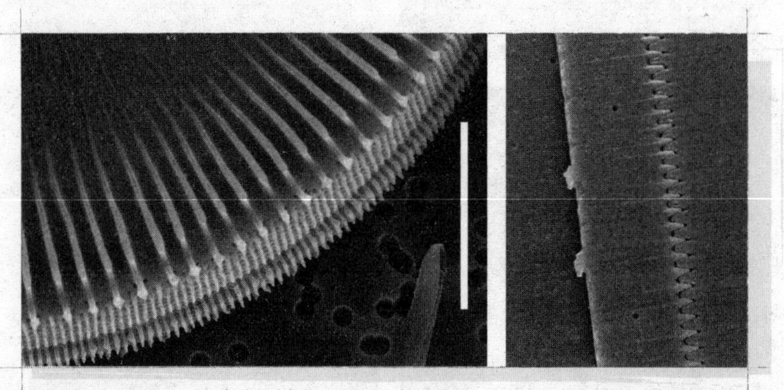

Figure 13.2 Close-up pictures showing the exquisite precision of frustule craftsmanship. The magnification is the same in both pictures, with the white bar indicating a length of ten millionths of a meter, or one hundredth of a millimeter.

by everyday opportunities to experience the kind of exhilaration I am talking about. I recall endless fascination as a child watching houseflies frolic and groom, ants march along their invisible trails, and pill bugs roll themselves into protective balls until danger had passed, wondering how such small creatures could be so complete in their animal behavior. So learn from the children. If the exhilaration is gone, I guarantee the problem isn't that wonders have ceased or that your advanced understanding has emptied them of all mystery.

Don't let the Internet replace your firsthand experience of life, but do let it extend your experience. Having dived in, take another look at fish, for example. And I mean just *look* at them, without trying to figure them out or classify them. On second thought, do classify them, but more as a moviegoer than an ichthyologist, and let the exercise provoke you to wonder. How on earth do we find ourselves on a planet where the great emotive categories of film, story, and stage are so beautifully represented by *fish,* of all things? Are you in the mood for fantasy? Try Merlet's scorpionfish or the mandarinfish (see Plate 3). More inclined toward drama or romance? You can't beat the well-known Siamese fighting fish. How about comedy? My personal favorites are the fringehead fish and the red-lipped batfish, though the options are numerous. Horror? Lots of possibilities here as well. My picks would be the giant stargazer and the fangtooth. And of course we've already considered the mighty salmon as the master of epic tragedy.

What's going on here? Why this strong resonance between the physical appearance of fishes and our own emotional makeup? And why is this so typical of life? We can't help noticing not just that life comes in a staggering variety of forms but also that so many of these forms strike a chord deep within us, as though they were *meant* to do just that. So for us to conclude merely that each distinct form of life had to be invented would be to grasp something with the head only to miss something bigger with the heart. In life we have proof not just that a great Inventor exists but, more significantly, that a great *Creator* exists—someone who invested not just intellectually but also emotionally, just as we invest in our creations.

Closure

The first world, then, seems to be not the world of mere *thinkers* but the world of *persons,* complete with personhood and personalities. Certainly this is the richest world, not just comparatively but *categorically,* which makes nonsense of any thought that a lesser one brought this greater one into being.

When we take this to be the right picture of reality, the conclusions we have drawn become even more potent. We decided that insight and purpose are essential ingredients for invention, and in this way we distinguished the intentional from the accidental, all the while sensing this distinction to be of profound significance. Now we see just how profound it is. Inventors are not just inventors but *creators,* and creators are not just creators but *persons.* And however insistent the materialists have been that a person is nothing more than a special arrangement of ingredients from the periodic table, we now know with categorical certainty that this cannot be. Our own thinking can't possibly be reduced to any physical process because it collapses the moment we suppose otherwise. We are persons, dwellers in the richest of all worlds, and this rich world of personhood we each inhabit *had* to come from a source where riches of this kind are well known.

We have our answer, then, and we can thank the children for declaring it. The source from which everything else came is not a *what* but a *who*. Of the millions of species participating in this remarkable adventure called life, only ours has been given the ability to grasp this most crucial truth. We can and we do, from an early age. Perhaps we also *should* grasp it, and having grasped it, perhaps we should hold on to it. Maybe we should pay less heed to the internal tensions that would pry this truth from us and more to the truth itself. This truth does, after all, have every appearance of being *good,* so maybe we resist it for no good reason. If personhood is at the very center of reality, and if the resonance between us and our Creator is as deep as we have seen, then friendship with him can't be far off.

Who knows? He might even understand our cosmic authority problem.

CHAPTER 14

THE NEW SCHOOL

Reflection is valuable near the end of a successful journey. Thinking back, we started with what seemed like very little. We had nothing but the unresolved question of where we came from and the determination to follow the truth to the answer. The problem wasn't that we had no answer but rather that we had *two* answers that contradicted each other. In our childhood, if not since, our design intuition assured us that life could only be the handiwork of God, or someone like him. As universal as this intuition is, though, it is almost universally opposed by the technical experts on life. None of us have been able to erase the intuition, but many of us have struggled to defend it against this professional opposition—or even to know whether it *ought* to be defended.

Summoning our courage, we set out to see whether there might be more to this humble intuition than meets the eye. Everything in our experience told us this had to be so. Some things really *are* too good to happen by accident, and if experience affirms this principle even for mundane things like bricks and shoes, how can exquisite things like spiders and orcas be exceptions?

Having now added to our experience this journey we've nearly completed, we see that—far from being exceptions—living things are the most striking examples of this principle. No high-level function is ever accomplished without someone thinking up a special arrangement of things and circumstances for that very purpose and then putting those thoughts into action. The hallmark of all these special arrangements is high-level functional coherence, which we now know comes only by insight—never by coincidence.

This vindication of our design intuition has been one of the major accomplishments along our journey, but there have been others. Having been reminded of how human scientists are, we've learned to let go of the utopian version of science, which never resembled real science anyway. Likewise, with the affirmation of our design intuition came the realization that scientific thinking is part of what we all naturally do. The community of professional scientists gets some things right and some things wrong, just as every other community does. All humans are scientists, and all scientists are human.

Topping off these accomplishments is the weighty realization that the great Cause of everything clearly reveals himself not as an impersonal force but as a very personal God. Even *this* fits with the universal design intuition. Creation is only ever accomplished by drawing upon what exists, and personhood, so fundamental to our existence, must therefore have come from someone in whom it already existed. Persons can only have come from a personal God.

We had the right answer to our big question back in our childhood, though for many of us it was misplaced somewhere between then and now. Thankfully, what was lost has now been found: We owe our existence to the personal God who understands our existence. We were never alone.

Nothing I say in the remaining pages will match the importance of this rediscovered truth, so I won't even try for that. I hope instead to help us start to think about the *breadth* of its importance. When I said at the start that *who we are* has everything to do with *how we ought to live,* I didn't mean this in a moralistic sense. Certainly, the nihilistic message that David Barash preaches to his students is false, but as we've begun to see, God's presence and the deep connection we have with him through personhood has much more exciting implications than the reality of right and wrong.

To give us a taste of this, I want to end by giving us a glimpse of how exciting the transformation of biology would be if a true understanding of its place within the big picture were to take hold. And if biology can be transformed in this way, why not other pursuits as well? Laying claim to these prizes will require much hard work, but in the few remaining pages I hope at least to convince you that prizes of this transformative kind are really out there waiting to be claimed—well worth the labor this will require.

We'll start by thinking not about biology but about another field—one that was bursting with the excitement of newness not so long ago.

MIND OVER MATTER

In the mid-1930s the foundations of an entire discipline took shape in the mind of a young Englishman by the name of Alan Turing. Calculating machines had been conceived and built earlier, but Turing's invention was altogether different and better. Where others had invented interesting *things*, he invented the interesting *idea* that suddenly made everything click. His conceptual machine, immortalized as the *Turing machine*, became the defining model for the programmable calculating machines we know as *computers* (Figure 14.1).

Turing hit the mark so perfectly that everything else, including much of our common knowledge about computers, turns out to be incidental. We think of computers as electronic

Figure 14.1 A Turing machine (conceptual) that computes by inverting pennies arranged in a long row. Anything that can be switched between two distinguishable physical states can be used instead of pennies, which is why we think of computation in terms of symbols (zeros and ones) instead of physical states. From an initial start-up state, the machine "reads" the penny under the pointer and then does whatever the corresponding action rule says. Actions may include inverting that penny and/or changing the internal state before moving to the next penny, either left or right as specified by the rule. In essence, the row of pennies functions as memory for input and output, and for performing the computation. The machine is like the CPU of a modern computer; it has a fixed architecture that can be put into any of a large number of temporary internal states, after which it automatically moves itself from state to state by applying its built-in action rules to the whole state (including the state of the penny being read). Although the vast majority of possible Turing machines do nothing interesting, a special subset known as *universal Turing machines* are capable of performing *any* algorithmic computation, provided they are "programmed" with a row of pennies carrying the necessary information, along with enough additional pennies for adequate working memory. Like all interesting inventions, universal Turing machines require extensive functional coherence, making them fantastically rare within the immense space of possible Turing machines.[1]

devices with keyboards and displays on the outside and silicon chips on the inside because in our experience this is what they are. But these familiar objects are only *one* way of giving physical form to a Turing machine. In fact, long before Turing was born, Charles Babbage designed digital computing machines with rotating gears and cylinders. Such things are long forgotten, but interest in modes of computing having nothing to do with silicon chips lives on. Bringing sense to all of this is Turing's insightful way of understanding the essential elements common to *all* forms of digital computation.

We've become dependent on chip-based computing in recent decades, but we've always been more directly dependent on another kind of computing, this being the kind that occurs inside our brains. I'm referring not to mental calculation but rather to the physical processes carried out by the gray matter inside our skulls. For example, for the light that enters our eyes to be translated into a conscious visual scene, it must first be processed by an extraordinarily sophisticated network of neurons in the occipital lobe at the back of our heads. You might suppose the secrets of this signal processing have yielded to modern brain research, but the truth is that the details remain completely mysterious. "In the deepest sense, we do not know how information is processed, stored, or recalled" in the brain, said the experts at a recent workshop on brain function.[2]

The staggering complexity of the brain's structure, with its hundred trillion neural connections, is certainly one reason for the slow progress, but I have to think that false preconceptions are another. Materialism, in particular, has constrained thinking within brain science as severely as it has elsewhere. Even the title of that workshop—*From Molecules to Minds*—is a proclamation of the view that mental processes are grounded in molecular processes.

Many outstanding experts see through this. For example, Jeffrey Schwartz, a scientist in the Department of Psychiatry at UCLA School of Medicine, has written a book—*The Mind and the Brain*—that "challenges the idea that we are merely biologically programmed automatons and proves that we have the [mental] power to shape our brains."[3] People like Schwartz, make me think it's possible for the materialistic view to be displaced from its underserved position of authority.

At the same time, I can see how people get caught up in the materialistic view of the mind. Even when computer technology was in its infancy, the idea of computers being "thinking machines" had a certain seductive appeal. They *do* seem to think, at least in the sense of arriving at answers we arrive at only after much thought. Since brains are physical things, should we doubt they are thinking things as well, making thought more sophisticated and less mechanical than computation, but no less physical?

However plausible this may appear when approached from that angle, the previous chapter's thought experiment exposes the underlying fallacy. The plain fact is that we consciously ground our thoughts in conceptual realities, not in physical realities. So to claim that the underlying reality of our thought process is *physical* is to claim that what is actually happening when we think is profoundly different from what we *think* is happening when we think. Since the *conceptual* realm is where thinking must occur, if it is to occur at all, the proposition that this realm is not fundamentally real would, if taken seriously, force us to abandon reason as a hopeless pursuit.

Thankfully, we have a much more satisfactory alternative. When we accept the fundamental reality of the conceptual realm, we see that computers don't really think after all. Like can openers and mousetraps and robotic pool cleaners, they give the *impression* of knowing what they're doing only because their inventors, who really did know what they were doing, imparted genuine cleverness to their designs. As we saw in chapter 9, the hierarchical structures of these inventions reveal how their inventors thought. Far from being exceptions to this principle, computers and the applications they run are striking examples of it.

The human brain is different, though. Being the most remarkable component of the human body, it is arguably the most outstanding physical invention ever to exist. Even more spectacularly, the brain is the one physical invention that serves as an interface between the physical and conceptual worlds.

Take a moment to let the significance of this send shivers down your spine. This universe has within its vast swirling wisps of scattered elements a fixed number of *connecting points* between the immense realm of things and the infinite realm of thoughts. You know this because one of these connecting points is humming with activity right now, inside your skull, enabling you to reconstruct thoughts from physical symbols on paper or on an electronic display.

In purely material terms, these connecting points are as nothing—vastly outnumbered, outweighed, and outpowered by the stars in our galaxy, which is only one of a hundred billion galaxies. But that assessment flips the moment we take all of reality into account. Significance isn't measured in kilograms or light years because, like truth, it belongs to the realm of *ideas.*

Significance is therefore weighed only by those capable of weighing ideas. Once we recognize this, the profound importance of these special locations in our universe becomes apparent. These connecting points are the places—the *only* places—where the world of atoms and the world of ideas are made to shake hands. Poet meets muse. Sculptor touches stone. Melody finds strings. Ideas flow to paper. Thirst is quenched. Loneliness ended.

Absolutely everything of momentous importance in our universe is occurring at these special points, which is why we name and cherish the possessor of each one—why we celebrate their births and mourn their passing. If the galaxies out there were capable of grasping the meaning of the universe, their attention would be fixed on one little planet circling an ordinary star situated in a minor arm of an otherwise ordinary spiral galaxy. What is present on that one little planet makes this particular galaxy—the one named after milk—utterly extraordinary.

Two Roads

A sobering realization presents itself at this point, followed by an exciting one. If materialism continues to dominate the sciences, then brain research will continue to be driven by the pursuit of something unreal, namely the molecular basis of mind. In that case we will have missed a huge opportunity to learn about what *is* real. Of course, any number of facts and physical details will be discovered and cataloged as we continue down this old road, but the full significance of those facts can't possibly be grasped while we labor under a false understanding of what a brain is. On the other hand, if the blinders could somehow be removed and this whole caravan of effort redirected to a *new* road—one that starts with simple truths about life and follows these to technical truths—might we see huge advances in a short time? Might we even find that the facts that have already been cataloged add up to something much bigger and more coherent when interpreted through the right lens? The possibility is intriguing (Figure 14.2).

And what about the rest of biology? After all, the old materialist road has a great many side paths branching off, where specific aspects of life are studied. If we imagine the new road, some distance away, we see similar branching paths. Signs marked Cyanobacteria, and Diatoms, and Proteins, and thousands of other subjects line both main roads. Scientists who access their work from either of these two roads would be asking many of the same questions and using mostly the same methods to answer those questions. The search for new diatom species and the methods for visualizing their intricate outer casings, for example, would be identical in the two communities. And yet we anticipate major differences on the paths marked Brain Function.

What accounts for the major differences?

For one thing, biologists who have traveled the materialist road to their specializations never ask *why* things are as they are, at least not in the deepest sense of the word. If diatom and orca and human are nothing more than leaves drifting on water, dispersed from a common starting place by fluctuating currents, then questions of *purpose* are misplaced. We can talk about *how* this leaf came to be here and that leaf there, but since matters of *how* generally borrow their significance from matters of *why*, this isn't a particularly inspiring line of inquiry. To add another dollop of dull, Darwinian answers to the *how* question all start to sound pretty similar after you've heard a few of them. As far as I can tell, the only reason for excitement on the path marked Evolution is that no one really *can* view life in this ho-

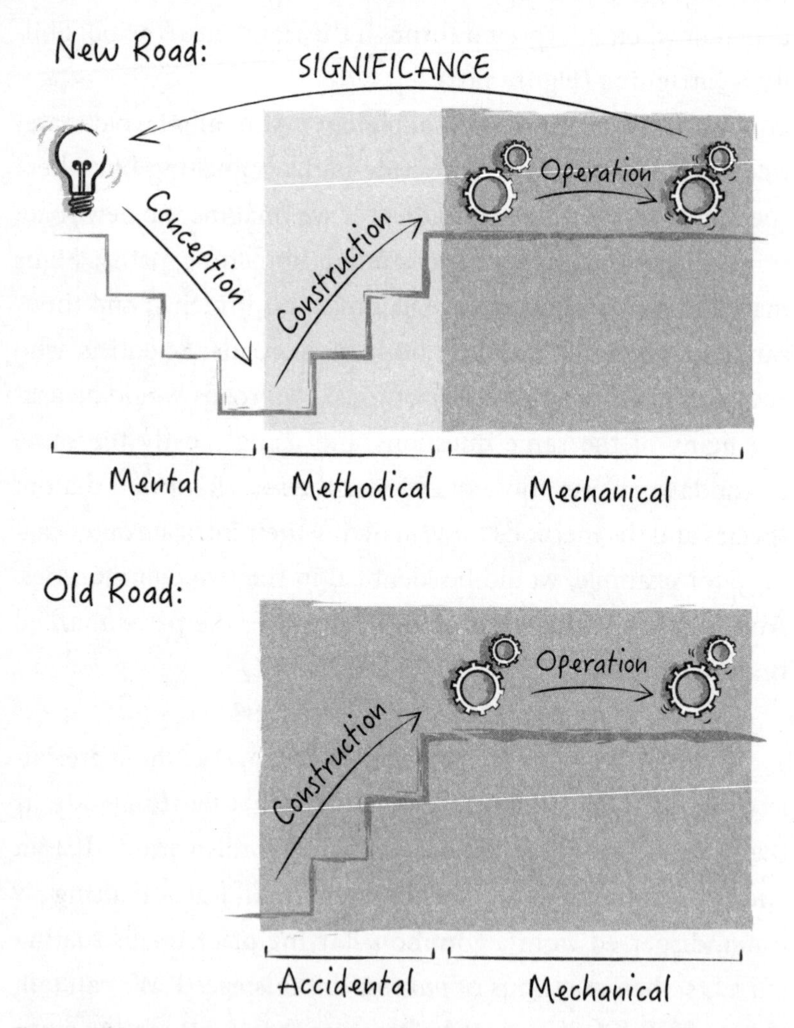

Figure 14.2 New-road and old-road interpretations of Figure 9.1. By denying all conceptual aspects of biological invention, the materialist view cuts off consideration of an invention's *significance*. With significance off the table, and purpose with it, there isn't much to be said about an invention beyond how it works and the effects it has.

hum way. Life is so arrestingly compelling that *it* becomes the spice in what would otherwise be a depressingly bland dish.

WINE WITHOUT COST

This profound disconnect between a Darwinian understanding of life and a true understanding deserves more attention than I can give it here. I'm reminded of the Dawkins-sponsored ads that appeared on the sides of London buses, declaring in bright colors, "There's probably no God. Now stop worrying and enjoy your life" (Figure 14.3). Interesting logic. You are a meaningless by-product of the percolating ooze in some ancient pond, soon to return to the dead chemicals that burped your ancestors out, sooo . . . go enjoy your life! As strange as this advice seems to people who don't see things the way Dawkins does, it begins to make sense if you pretend that the enjoyment of good things carries no obligation to appreciate their source. For most of us there *is* much to enjoy in life, and moreover this bounty is so familiar that many of us take it for granted. Having done just that, atheists of the Dawkins variety feel free to pair this wine of pleasure with whatever dish of explanation they find most convenient, no matter how grotesquely incompatible the two may be. *You and I* may be repelled by the food on their plates, having reasoned that it robs life of all hope and meaning, but that's only because we take these implications far more seriously than they do. For them the plate is just an excuse to keep getting refills on the wine.

I have a great deal more respect for Nagel's version of atheism, where the desire to drink the wine on one's own terms is acknowledged, but not without a recognized obligation to live off the food on one's own plate. To take this obligation seriously is both admirable and admirably risky—in the end there may not *be* a dish that justifies taking the wine on one's own terms.

The situation in biology parallels the Dawkins disconnect, I think. All biologists are impressed by life. If they weren't, they wouldn't have devoted themselves to studying it. The problem is that life's splendor is so obvious they tend to take it for granted, pushing it to the background to make room for their academic theorizing. But instead of following the implications of their theories to their logical ends, biologists compartmentalize. *Yes,* orcas are the product of blind material forces that had no ability to conceive them, and *yes,* they take our breath away whenever we watch them. Never mind how these two affirmations fit together. Just pretend they do. Life is meaningless. Isn't it beautiful?

Figure 14.3 Richard Dawkins posing with ad creator Ariane Sherine in front of one of the London buses carrying the "peaceful and upbeat" message.

TWO SCHOOLS

Only a very few research scientists have the opportunity to work against that disjointed view by openly studying life as something clearly and cleverly designed. I am one, and I can count the others on my fingers.[4] There are more who would *like* to have this opportunity, as shown every now and then by a paper that gets past the policing system of an establishment science journal. A recent example is a description of the architecture of the human hand as being "the proper design by the Creator to perform a multitude of daily tasks in a comfortable way."[5] Infractions like this almost always bring out the whistle-blowers, which almost always brings a reprimand. Everyone must decide for himself or herself what they can do under the shadow of the materialist flag, knowing that if they press too hard they may lose even the small opportunities they once had.

Real case histories bear this out, time and again. Two months after this paper on the human hand was published, it was retracted—not by the authors but by the journal (*PLoS One*), and not for any technical error that could be described, but because of "concerns with the scientific rationale, presentation and language."[6] This sounds very much like the vague language of prejudice. Consistent with that, the retraction mentions only one specific objection: "Following publication, readers raised concerns about language in the article that makes references to a 'Creator.'"[7] Evidently *PLoS One* marches on command whenever a whistle is blown.

Now, if there were a well-known and consistently applied rule forbidding references to deity in science papers, then I suppose actions of this kind would be excusable. Instead there is a glaring double standard. A decade ago, seventeen pages of the journal *Gene* were devoted to a rant against intelligent design that had *plenty* to say about God—all negative, of course. In the peer-reviewed pages of this establishment journal, Emile Zuckerkandl was allowed to speak his mind with no interference from the thought police. "Kronos is a God who cannot be denied by any other God. Nor was he by the God of the Jews . . . no God can be almighty."[8] Ten years on, this one still hasn't been retracted. Maybe nobody raised concerns.

I suppose I could have turned this book into a sustained protest against the culture that gives rise to all this injustice. That's not the book I felt had to be written, though. Instead, my aim has been to show you that there's a much more compelling view of life than the materialist view and that this compelling view also happens to be *innate*—known by us from early childhood and stubbornly persistent thereafter, such that to deny it requires sustained effort. To achieve my aim I've focused first on life generally and then on humanity specifically, hoping that when people see how durable the simple truths about all life and about human life are, they will be inspired to search out those further truths we need for an adequate grasp of reality. In other words, I hope to have begun the process of inquiry, not to have completed it.

With respect to life generally, widespread participation in this ongoing search would give new vitality to the study of life. In the first place, those *why* questions that have no place on the old road become the intellectual core of the new road. Until this road becomes more populated, though, some imagination will be needed to picture the way things could be. One way to get a feel for biology's new road is by comparing an engineering discipline, where *whys* are a staple, with an imaginary *old-road* version of itself. To do this, imagine that all of humanity has suddenly suffered a highly selective form of amnesia that erased all knowledge of computers. To top off this pretend disaster, suppose all documented knowledge about computers vanished—everything from websites to textbooks to videos. A moment ago humans had a deep understanding of computers, but now we find ourselves in a state of ignorance, marveling at these complex machines and wondering how they work.

As more and more technically minded people start examining these mysterious electronic devices, two schools of thought begin to emerge. The old school (on the old road) adopts the view that computers happened by accident, through a happy but unintended convergence of circumstances, whereas the new school appeals to the universal design intuition, arguing that because computers show all the hallmarks of inventions, they must have been invented. Students of both schools initially spend their time observing what computers do before moving toward experimentation, where they try to understand how the various parts enable them to do what they do.

This is where we begin to see the different schools leading their students in two very different directions. Students in an old-school computer science lab find themselves surrounded by dusty displays of half-dissected computers with faded labels naming the major parts. The place of honor at the front of the room is given to a most impressive display: a collection of microchips arranged according to the number of legs they have, each carefully impaled with a pin and identified by a handwritten Latin name. Working in pairs, the young computer scientists pry little pieces off boards taken from dead computers, carefully sketching them in their laboratory notebooks, knowing they will have to memorize the technical names and locations of each for Friday's test. The big research universities are abuzz with much more advanced work, of course. Thermal imaging is used to see how much heat the various computer parts produce in real time and how this depends on the application being run. These big-time scientists can even read the zeros and ones from an entire hard disk and test the effects of changing a zero to a one, or vice versa. All very high tech.

Nevertheless what eludes all these old-school computer scientists, despite their hard work, is the hefty matter of understanding what a computer *is*. To know what a computer is made of and what kinds of things it can do is one thing. To know what it *is*, is another. The first is of some value, but that value will be severely limited without the second. Had the young Alan Turing been brought up in this imagined world, much would have been lost. All thinkers are given a context in which to think, and when part of that context is the unquestioned assumption that the things being studied were caused only by other *things*, then the *ideas* that may have been the real cause are bound to be overlooked.

Shortsightedness of this kind begins with a failure of conviction. When we abandon our design intuition we lose the most potent alarm that would tell us the wrong road has been taken. Having silenced that alarm, workers on the old road may continue for any number of generations, assuring themselves of their productivity by pointing to the continual growth of knowledge, without ever pausing to contemplate the distinction between knowledge and understanding. Ironically, the inadequacy of the old-school perspective perpetuates the old-school work by guaranteeing that the goal of complete knowledge will never be reached.

The old road has no end.

GLIMPSING THE NEW

I have a strong feeling the new road has no end either, but here the lack of an end is a very good thing. We see hints of the new road even now, as though occasionally a scientist takes a side path from the old road so far that they come within a stone's throw of the new road, maybe without realizing it. I think of Princeton physicist William Bialek, who heads a research team that measures how well various biological functions are performed relative to the lofty standard of physical perfection. He says:

> Strikingly, when we do this (and there are not so many cases where it has been done!), the performance of biological systems often approaches some limits set by basic physical principles. While it is popular to view biological mechanisms as an historical record of evolutionary and developmental compromises, these observations on functional performance point toward a very different view of life as having selected a set of near optimal mechanisms for its most crucial tasks. . . . The idea of performance near the physical limits crosses many levels of biological organization, from single molecules to cells to perception and learning in the brain, and I have tried to contribute to this whole range of problems.[9]

In other words, in design situations where human engineers would want to push the limits of physical possibility if they could, we often find that biological systems operate at or near those limits.

There's more subtlety to this claim than I can unpack in a few words, and some may be inclined to dismiss it for that reason. To fully grasp the point, you have to look quite deeply into actual design constraints and objectives. For example, gazelle legs don't propel gazelles to speeds remotely approaching the speed of light (the absolute physical speed limit), but then neither would human engineers set out to make an all-terrain vehicle that moves anywhere near that speed. On the other hand, the eyes of cats *do* approach the physical limit of single-photon sensitivity, and the antennae of certain male moths *do* achieve single-molecule sensitivity to sex pheromones, and certain enzymes *do* approach the physical limit of proficiency—processing their reactant molecules as fast as diffusion can deliver them. To anyone with an appreciation of design challenges, the long list of believe-it-or-not facts like these coming from biology is truly astounding.

Since all these facts came out of the old school, firmly situated on the old road, you may be wondering about the benefit of relocating biology to a new school on the new road. Here I go back to our mental picture. The greatest loss suffered by our imaginary old-school computer scientists is not a shortage of observations made as outsiders peering into their subject but rather their having excluded themselves from becoming *insiders*—from grasping their subject deeply enough to become *participants* in it. Now, by suggesting that the materialist commitment has likewise excluded biologists from participating in their discipline, I don't mean biologists would be designing and building new life forms if Darwin hadn't taken us down the wrong road.[10] I mean that biologists ought by now to have grasped what life *is* with enough clarity to inspire a much deeper intellectual appreciation of life. That insight, completely missing from today's biology, would shed new light on every subdiscipline. For something so fundamental not to have this comprehensive effect is inconceivable.[11]

Consider popular wisdom about genes and DNA. Just as most people think scientists have figured out how the brain works, so too they think scientists have figured out how DNA works. By my casual observation, most nonscientists—and some scientists as well—think the blueprint from which every living organism was formed is written on that individual's genome in the language of genes. Accordingly, geese honk because they have the honk gene, and hyperactive dogs yap because they have the hyperactive-dog gene. Likewise, by this popular view people who can sing or whistle received these abilities by receiving the corresponding genes. The master template for specifying *all* our attributes became public with the publishing of the human genome, supposedly, so all that remains is to finish the task of assigning traits to genes and to empower every person to read and interpret his or her own personal blueprint.

Coming from that viewpoint, most of us would be shocked to know the actual state of ignorance with respect to DNA. The view that most aspects of living things can be attributed neatly to specific genes has been known by geneticists to be false for a long time, this being the first common DNA myth to fall. A second, which has fallen only quite recently, is that scientists even have a clear understanding of what a gene *is*. Without exaggeration, a recent article in *Science and Education* stated that "the gene concept is currently in crisis."[12] It turns out that the simple picture of a gene as a section of DNA that encodes a protein, as described in chapter 3, no longer holds for anything but bacteria. To give you an idea of how far current thinking has moved from that simple view in recent years, consider this excerpt from a prominent article in *Genome Research:*

> One metaphor that is increasingly popular for describing genes is to think of them in terms of subroutines in a huge operating system (OS). That is, insofar as the nucleotides of the genome are put together into a code that is executed through the process of transcription and translation, the genome can be thought of as an operating system for a living being. Genes are then individual subroutines in this overall system that are repetitively called in the process of transcription.[13]

The fact that ideas like this or like those of William Bialek can be expressed under the materialist flag is a good thing. The problem comes when people want to take such radical thoughts seriously. For example, if genomes really are like operating systems, then the thought of them carrying the blueprints for building the bodies of their possessors is as wrong as the thought of the iPhone operating system carrying the plans for manufacturing the iPhone itself. And if we allow ourselves to take *that* idea seriously, then the thought of genetic mutations having changed a primordial organism into all modern forms of life is seen to be confused *over and above* its mistaken reliance on accidental causes. For an iPhone 5 to be converted into an iPhone 6 by upgrading its operating system is *categorically* impossible—with or without insight. Extending that principle to life would take us beyond our conclusion that modern life can't be the product of *accidental* mutations—implying it can't be the product of mutations *at all*.

So if this is where the thoughts go, are we allowed to go with them?

My point—my *plea*—is that scientists ought to be encouraged not only to develop ideas that touch biology so deeply but also to take those ideas seriously enough to test and extend them. Efforts of this kind ought to be hailed as the surest sign that the scientific community is alive and well. If we can agree on that, then there are bright days ahead. Indeed, I am convinced that the very best days in the study of life were not the days that catapulted that laboratory under the direction of Max Perutz to lasting fame—the days when life's smallest pieces were revealed to humanity for the first time. The very best days, still to come, will be those when all the pieces come together under a set of organizing principles by which they suddenly make sense.

Biology awaits its Turing machine.

COMMON GOOD

That the deepest questions in biology have not yet been answered means they are still asking to be answered. Anyone who cares to examine the facts carefully will see that the old answers were wrong. They have now been erased, in our minds anyway, and we must sit down to take the test again, with new minds and new resolve. Having learned much since Darwin's day, we have every reason for optimism this time. Speaking as a scientist, I can't think of a more attractive message to convey to young people of technical ability.

Speaking as a human, though, I see something even more beautiful. Yes, the deepest questions in the scientific study of life are up for grabs, and this is exciting for the technically minded. But the deepest truths of life itself, and of human life in particular, were never really up for grabs. These were never restricted to the most clever. Some things, of course, can only be seen by standing on the shoulders of giants, but the most crucial things have always been seen best by standing on the ground.

ACKNOWLEDGMENTS

A big *thank you* to the many people who've helped me in various ways. Thanks first to my agent, Giles Anderson, for his work at the early stages to get the book underway. My editor, Katy Hamilton, oversaw the whole project at HarperOne and, with executive editor Mickey Maudlin, recommended structural changes to the manuscript that turned out to be extremely helpful. Thank you both! Noël Chrisman's attention to detail improved the book considerably at the copyedit stage, for which I'm very grateful. And thanks to Ann Edwards, Anna Paustenbach, and Jane Chong of HarperOne and to Rob Crowther of Discovery Institute for the care and insight they brought to the task of increasing the book's visibility.

I thank Brian Gage for directing the illustration work and Anca Sandu for adding her creative talent to the project (along with her patience). Thanks also to Rachel Aldrich for her work on image acquisition.

I deeply appreciate the time many people took to read the manuscript, in part or in whole, and to offer their comments. Titus Kennedy, Casey Luskin, George Montañez, Steve Zelt, Steve Fuller, Bill Dembski, Jonathan Wells, Rebecca Keller,

Mariclair Reeves, Jacob Koch, Grant Gates, Ann Gauger, Fraser Ratzlaff, Chuck Wallace, and Eric Garcia are all to be thanked for this, as are Jim and Paula Thomas for generously giving me a beautiful place to read the manuscript myself.

I'm very grateful to Jim Wiggins for supporting this project, and to the Biologic board for giving me the time to complete it: Steve Meyer, Ted Robinson, Scott Webster and former board member Chuck Anderson—a friend who didn't live to see the book published but who was a huge inspiration along the way.

Special thanks to my brother Ron for his ongoing interest in this work and to my father, Rick, for his unfailing and enthusiastic support and inspiration.

Among the young people who've been an inspiration to me as I've thought about the design intuition are my niece, Jordan, and my own three children—Noelle, Caitlyn, and Ryan (credit for the *Undeniable* title going to Caitlyn, by the way).

A "gazillion" thanks to Anita—my wife and best friend—for gently steering me when I've needed it most.

And finally, quoting my Ph.D. thesis: "I should express my gratitude to and dependence upon the One who made both the universe and the mind with which I seek to comprehend it. I have accomplished nothing of significance apart from Him, nor will I ever."

NOTES

Chapter 1: The Big Question

1. Francis Crick, *What Mad Pursuit: A Personal View of Scientific Discovery* (New York: Basic Books, 1988), 138.

2. All editions of Darwin's book are freely available online. The Darwin Online site is the best resource: http://darwin-online.org.uk/EditorialIntroductions/Freeman_OntheOriginofSpecies.html.

3. Charles R. Darwin, *The Origin of Species by Means of Natural Selection, or the Preservation of Favoured Races in the Struggle for Life,* 6th ed. (London: John Murray, 1872), 424.

4. Thomas Nagel, *Mind and Cosmos: Why the Materialist Neo-Darwinian Conception of Nature Is Almost Certainly False* (Oxford: Oxford Univ. Press, 2012).

5. There are clear signs that the days of the materialist flag are numbered. According to Robert Koons (University of Texas at Austin) and George Bealer (Yale University): "Materialism is waning in a number of significant respects—one of which is the ever-growing number of major philosophers who reject materialism or at least have strong sympathies with anti-materialist views." Robert C. Koons and George Bealer, eds., introduction to *The Waning of Materialism* (Oxford: Oxford Univ. Press, 2010).

6. Thomas Nagel, *The Last Word* (Oxford: Oxford Univ. Press, 1997), 130–31.

Chapter 2: The Conflict Within

1. Darwin to Joseph Hooker, 1871, as recorded in a footnote in *The Life and Letters of Charles Darwin, Including an Autobiographical Chapter,* ed. Francis Darwin, vol. 3 (London: John Murray, 1887), 18.

2. Alison Gopnik, "See Jane Evolve: Picture Books Explain Darwin," Mind and Matter, *Wall Street Journal,* April 18, 2014, http://online.wsj.com/news/articles/SB10001424052702304311204579505574046805070.

3. A *teleological* explanation is an explanation in terms of purpose. The source of the quote is Art Jahnke, "The Natural Design Default: Why Even the Best-Trained Scientists Should Think Twice," *Bostonia,* Winter/Spring 2013, www.bu.edu/bostonia/winter-spring13/the-natural-design-default/.

4. Plutarch, "Fortune," trans. Frank Cole Babbitt, in *Moralia,* vol. 2, *Loeb Classical Library* (Cambridge: Harvard Univ. Press, 1928), 87.

Chapter 3: Science in the Real World

1. "Mathematical Challenges to the Neo-Darwinian Interpretation of Evolution," *The Wistar Institute Symposium Monograph Number 5,* ed. P. S. Moorhead and M. M. Kaplan (Philadelphia: Wistar Institute Press, 1967).

2. Michael Denton, *Evolution: A Theory in Crisis* (London: Burnett Books, 1985). Other significant books during this period include C. Thaxton, R. L. Olsen, and W. L. Bradley *The Mystery of Life's Origin: Reassessing Current Theories* (Dallas: Lewis and Stanley, 1984) and A. E. Wilder-Smith *The Natural Sciences Know Nothing of Evolution* (Master Books, 1981).

3. Scientists call the appendages *side chains,* which is confusing because most of them don't look like chains, whereas the proteins made by connecting amino acids *do* look like chains. I will stick with *appendages* for that reason.

4. Some proteins need help during folding to avoid interference from the many other proteins that crowd the cell interior. Cells use special proteins called *molecular chaperones* to assist folding, some of which form specialized compartments in which new protein chains can fold without interference. The compartment-forming chaperones are called *chaperonins.*

5. Because any of the four bases (A, C, G, and T) can be placed at any base position, the number of possible sequences in a run of consecutive bases is calculated by multiplying by 4 repeatedly—one 4 for each base position. This means a run of two bases allows 16 possibilities ($4 \times 4 = 16$), which isn't enough to specify each of the 20 amino acids. Life therefore uses codons of three consecutive bases, bringing the number of possibilities to 64 ($4 \times 4 \times 4 = 64$).

6. Denton, *Evolution,* 327.

7. Denton, *Evolution,* 323.

8. D. D. Axe, N. W. Foster, and A. R. Fersht, "Active Barnase Variants with

Completely Random Hydrophobic Cores," *Proceedings of the National Academy of Sciences USA* 93 (1996): 5590–94.

9. D. D. Axe, "Extreme Functional Sensitivity to Conservative Amino Acid Changes on Enzyme Exteriors," *Journal of Molecular Biology* 301 (2000): 585–95.

10. Transcribed from "Standing Up in the Milky Way," *Cosmos: A Spacetime Odyssey,* Fox, aired March 9, 2014, www.dailymotion.com/video/x25wxcj_cosmos-a-spacetime-odyssey-episode-1-season-1-standing-up-in-the-milky-way_tv.

Chapter 4: Outside the Box

1. D. D. Axe, "Extreme Functional Sensitivity to Conservative Amino Acid Changes on Enzyme Exteriors," *Journal of Molecular Biology* 301 (2000): 585–95.

2. E. M. Brustad and F. H. Arnold, "Optimizing Non-natural Protein Function with Directed Evolution," *Current Opinion in Chemical Biology* 15 (2011): 201–10.

3. M. Altamirano et al., "Directed Evolution of New Catalytic Activity Using the α/β-Barrel Scaffold," *Nature* 403 (2000): 617–22.

4. M. Altamirano et al., "Retraction: Directed Evolution of New Catalytic Activity Using the α/β-Barrel Scaffold," *Nature* 417 (2002): 468.

5. "The God Lab: Advocates of Intelligent Design Have a New Strategy, and It Has Science at Its Centre," *New Scientist,* December 16, 2006, 8–11.

6. "Top School's Creationists Preach Value of Biblical Story over Evolution," Guardian, March 8, 2002, www.theguardian.com/uk/2002/mar/09/schools.religion.

7. "Creationism Row Reaches UK," New Scientist, March 14, 2002, www.newscientist.com/article/dn2045-creationism-row-reaches-the-uk.html#.VIJV976QTO9.

8. "More Than Three-Quarters of a Century after the Scopes Monkey Trial, Darwin's Opponents Aren't Even Thinking of Giving Up," Creation vs. Evolution, part 1, *Newsday,* March 11, 2002.

9. "Six Days of Creation: The Search of Evidence; A Widening Movement against Evolutionary Theory Seeks Scientific Support," Creation vs. Evolution, part 2, *Newsday,* March 12, 2002.

10. Personal e-mail from Douglas Axe to Bryn Nelson, Tuesday, March 5, 2002.

11. D. D. Axe, "Estimating the Prevalence of Protein Sequences Adopting Functional Enzyme Folds," *Journal of Molecular Biology* 341 (2004): 1295–315.

12. For an extended example of whistle-blowing, see Barbara Forrest and Paul R. Gross, *Creationism's Trojan Horse: The Wedge of Intelligent Design* (Oxford: Oxford Univ. Press, 2004). Forrest's research for this book led her to me in 2000 and again in 2001, at which point it became clear that she wanted me to say my 2000 *Journal of Molecular Biology* paper had no implications for ID. After I refused to give her a statement to that effect, she and Gross gave their readers their own assurance in a rather panicked tone: "There is *nothing* in that article—certainly nothing explicit by any stretch of the verb's meaning—to 'support' ID. No working molecular or cell biologist, among several colleagues we have consulted, has reported otherwise." See page 41 (emphasis in original).

13. David P. Barash, "God, Darwin and My College Biology Class," *New York Times,* September 27, 2014, www.nytimes.com/2014/09/28/opinion/sunday/god-darwin-and-my-college-biology-class.html?_r=0.

Chapter 5: A Dose of Common Science

1. Michael Denton, *Evolution: A Theory in Crisis* (London: Burnett Books, 1985).

2. Alison Gopnik, "See Jane Evolve: Picture Books Explain Darwin," Mind and Matter, *Wall Street Journal,* April 18, 2014, http://online.wsj.com/news/articles/SB10001424052702304311204579505574046805070.

3. E. M. Brustad and F. H. Arnold, "Optimizing Non-natural Protein Function with Directed Evolution," *Current Opinion in Chemical Biology* 15 (2011): 201–10.

Chapter 6: Life Is Good

1. Charles Darwin, *On the Origin of Species by Means of Natural Selection,* 1st ed., chap. 4 (London: John Murray, 1859), 84.

2. A. K. Gauger and D. D. Axe, "The Evolutionary Accessibility of New Enzyme Functions: A Case Study from the Biotin Pathway," *BIO-Complexity,* no. 1 (2011): 1–17.

3. M. A. Reeves, A. K. Gauger, and D. D. Axe, "Enzyme Families: Shared Evolutionary History or Shared Design? A Study of the GABA-Aminotransferase Family," *BIO-Complexity,* no. 4 (2014): 1–16.

4. Charles R. Marshall, "When Prior Belief Trumps Scholarship," *Science* 341 (2013): 1344.

5. Stephen C. Meyer, *Darwin's Doubt: The Explosive Origin of Animal Life and the Case for Intelligent Design* (San Francisco: HarperOne, 2013).

6. In chapter 10, we'll see that genomic data contradict Marshall's suggestion. I let this go here in order to make a different point.
7. O. Khersonsky and D. S. Tawfik, "Enzyme Promiscuity: A Mechanistic and Evolutionary Perspective," *Annual Review of Biochemistry* 79 (2010): 471–505.
8. R. Mukhopadhyay, "Close to a Miracle: Researchers Are Debating the Origins of Proteins, *ASBMB Today* 12, no. 9 (2013): 12–13.

Chapter 7: Waiting for Wonders

1. Associated Press, "Seahawks Fans Set Noise Mark," ESPN, December 3, 2013, http://espn.go.com/nfl/story/_/id/10071653/seattle-seahawks-fans-set-stadium-noise-record.
2. Richard Dawkins, *The Blind Watchmaker: Why the Evidence of Evolution Reveals a Universe Without Design* (London: Longman, 1986).
3. Graham Bell, *Selection: The Mechanism of Evolution,* 2nd ed. (Oxford: Oxford Univ. Press, 2008).
4. People who've done the math on natural selection have also come to a modest view of its power. For those interested, two key findings have led to this modest view. The first is that factors other than genetic fitness tend to dominate in determining which individuals pass their genes to subsequent generations. This is reflected in the fact that the effective sizes of natural populations (i.e., the sizes of ideal populations lacking these competing factors that would carry the same genetic diversity as the wild populations) tend to be much smaller than the actual populations (see, for example, T. F. Turner, J. P. Wares, and J. R. Gold, "Genetic Effective Size Is Three Orders of Magnitude Smaller Than Adult Census Size in an Abundant, Estuarine-Dependent Marine Fish," *Genetics* 162 [2002]: 1329–39). The second finding is that the likelihood of a newly generated beneficial mutation becoming established in an *ideal* population is only twice the fractional fitness advantage. Since these advantages tend to be very slight and since real populations are far from ideal, likelihoods for real populations turn out to be surprisingly small. Calculated as twice the fractional fitness advantage multiplied by the ratio of the effective population size to the actual population size (M. Kimura, "Diffusion Models in Population Genetics," *Journal of Applied Probability* 1 [1964]: 177–232), the likelihood of the first possessor of a new beneficial mutation passing that mutation to the entire species can easily be less than one in a million. This improbability is further compounded by the rarity of beneficial mutations in the first place. Worse still, these widely acknowledged problems are all overshadowed by an even larger problem that *isn't* widely acknowledged, namely the irrelevance of the beneficial mutations

that do happen from time to time to the invention of anything remarkable. This book focuses on this more fundamental problem.

5. D. D. Axe and A. K. Gauger, "Model and Laboratory Demonstrations That Evolutionary Optimization Works Well Only If Preceded by Invention: Selection Itself Is Not Inventive," *BIO-Complexity*, no. 2 (2015): 1–13.
6. D. D. Axe, "Estimating the Prevalence of Protein Sequences Adopting Functional Enzyme Folds," *Journal of Molecular Biology* 341 (2004): 1295–315.
7. Axe, "Estimating the Prevalence of Protein Sequences Adopting Functional Enzyme Folds."
8. Axe and Gauger, "Model and Laboratory Demonstrations."
9. A. K. Gauger et al., "Reductive Evolution Can Prevent Populations from Taking Simple Adaptive Paths to High Fitness," *BIO-Complexity*, no. 2 (2010): 1–9.
10. Gauger et al., "Reductive Evolution."

Chapter 8: Lost in Space

1. Choose "Select whole earth"; type "2000" in the "No. of points" box; click "Get random point(s)"; click "See it on map."
2. Strictly speaking, systematic egg-hunt searches become progressively less blind as the search continues because each wrong guess is crossed off the list of remaining possibilities. However, the advantage of systematic searching like that only becomes significant in cases where a substantial fraction of the possibilities can be tested. Since we're interested in cases where this isn't true, we won't need to distinguish systematic searching from blind searching.
3. I mean physical events divided down to the scale of atomic interactions. Larger physical events are far less numerous and are generally composed of these atomic-level events, so this is a very generous way to estimate the maximum number of physical events that might accomplish something of interest. The estimate starts with the fact that a chain of cause-and-effect involving atomic interactions can't propagate effects faster than the speed of light. The time for light to travel an atomic distance (one angstrom) therefore sets a lower limit on the time interval needed for a physical event on this scale. There have been about 10^{36} of these event intervals in the 14 billion-year history of the universe. The maximum number of physical events over the history of the universe is calculated by multiplying this number of event intervals (10^{36}) by the number of atoms (10^{80}), resulting in a number (10^{116}) that fills just under one and a half lines of text.

4. A 1-pixel search space has 16,777,216 color possibilities. Every time we increase our search space by 1 pixel, we must multiply by 16,777,216. A 3-by-5 pixel rectangle contains 15 pixels, so we're multiplying that first 16,777,216 by 16,777,216 fourteen times.

Chapter 9: The Art of Making Sense

1. Richard Dawkins, *The Blind Watchmaker: Why the Evidence of Evolution Reveals a Universe Without Design* (London: Longman, 1986).
2. NASA, *Apollo 13 Technical Air-to-Ground Voice Transcription* (Houston: Manned Spacecraft Center, April 1970), www.hq.nasa.gov/alsj/a13/AS13_TEC .PDF.
3. In effect, we used random typing to estimate the fraction of the search space covered by the overly generous target. Once we have that fraction, the principle of reciprocal scale tells us how many *blind* attempts (which need not be random) are needed for success to be expected.
4. Low pixel resolution was used to avoid exaggerating the color extension. The point is that we see the subject matter of photos even when the pixel resolution is minimal, not because the pixels are invisibly small but because they work together in a visually coherent way.
5. *Mathematica* is a symbolic mathematical computation program developed by Wolfram Research, available at www.wolfram.com/mathematica.

Chapter 10: Coming Alive

1. Thomas Nagel, *Mind and Cosmos: Why the Materialist Neo-Darwinian Conception of Nature Is Almost Certainly False* (Oxford: Oxford Univ. Press, 2012), 32.
2. For a beautiful video of the sliding movements in one cyanobacterial species, see "Oscillatoria in Motion" from the YouTube collection of Bruce Taylor (www .youtube.com/watch?v=IP4ir0wumpw).
3. Richard Platt and Stephen Biesty, *Stephen Biesty's Incredible Cross-Sections* (New York: Alfred A. Knopf, 1992).
4. For the technically inclined, I highly recommend this review article: I. Grotjohann and P. Fromme, "Structure of Cyanobacterial Photosystem I," *Photosynthesis Research* 85 (2005): 51–72.
5. In addition to sunlight, air, and water, this larger project calls for traces of various minerals that are present in all natural bodies of water.

6. The term *intrinsically disordered proteins* refers to a class of proteins that exploits a partially (or completely) unfolded state to perform certain biological functions. Specific functions always require a degree of amino-acid sequence specificity, though this may be lower for these less structured proteins. However, since all life is absolutely dependent on a large number of precisely folded proteins, these are my focus. The arguments and evidence I present with respect to folded proteins are unaffected by the existence of proteins that don't fold.

7. D. D. Axe, "Estimating the Prevalence of Protein Sequences Adopting Functional Enzyme Folds," *Journal of Molecular Biology* 341 (2004): 1295–315.

8. A successful blind search for a folded protein of this size isn't quite *fantastically* improbable, so we can't say that it's physically impossible. However, we can say that it's *biologically* impossible by Michael Denton's 1 in 10^{40} criterion. Since the invention of proteins is a biological problem, Denton's criterion is more relevant. It makes little difference, however, because accidental inventions requiring just two new proteins are impossible by both standards.

9. C. Beck et al., "The Diversity of Cyanobacterial Metabolism: Genome Analysis of Multiple Phototrophic Microorganisms," *BMC Genomics* 13 (2012): 56.

10. K. Khalturin et al., "More Than Just Orphans: Are Taxonomically Restricted Genes Important in Evolution?" *Trends in Genetics* 25, no. 9 (2009): 404–13.

11. L. Jaroszewski, "Exploration of Uncharted Regions of the Protein Universe," *PLoS Biology* 7, no. 9 (2009): e1000205.

12. Some readers may prefer to credit highly intelligent life from another planet with the invention of earthly life. The first problem with this, if it's an attempt to avoid God, is that these hypothetical super-geniuses would themselves have had to be invented by an intelligent creator. In the end, there's no way around the fact that the first intelligent planetary life must have been created by an intelligence outside the universe, which sounds very much like God. The second problem is the *categorical* impossibility of anyone but God creating minds (a point to be picked up in chapter 13). Having now given you my reasoning, I won't hesitate to name God as the intelligent designer from this point forward. I do, however, want to stress that I am speaking for myself, not for the community of people who affiliate with ID.

Chapter 11: Seeing and Believing

1. Paul Rosenberg, "God Is on the Ropes: The Brilliant New Science That Has Creationists and the Christian Right Terrified," *Salon*, January 3, 2015, www.salon.com/2015/01/03/god_is_on_the_ropes_the_brilliant_new_science_that_has_creationists_and_the_christian_right_terrified.

2. Quoted in Natalie Wolchover, "A New Physics Theory of Life," *Quanta,* January 22, 2014, www.quantamagazine.org/20140122-a-new-physics-theory-of-life/.

3. Ernst Haeckel, *The History of Creation,* vol. 1, Project Gutenberg, www.gutenberg.org/files/40472.

4. C. E. Dobell, *Antony van Leeuwenhoek and His "Little Animals"* (New York: Harcourt, Brace, 1932).

5. Richard Dawkins, *The Blind Watchmaker: Why the Evidence of Evolution Reveals a Universe Without Design* (New York: Penguin, 1988), 9. Emphasis in original.

6. Quoted by permission from a pre-edit version of J. M. Tour, "Why Is Everyone Here Lying?" *Inference: International Review of Science* 2, no. 2 (2016).

7. Dawkins, *The Blind Watchmaker,* 49. Emphasis in original.

8. *Discover,* February 2005.

9. The evolved function was the equality function, which compares two binary input numbers and returns a 1 at positions where they match and a 0 at positions where they don't match. See R. E. Lenski et al., "The Evolutionary Origin of Complex Features," *Nature* 423 (2003): 139–44.

10. The first of the following papers is an analysis of the demonstration featured in *Discover magazine.* W. Ewert, W. A. Dembski, and R. J. Marks II, "Evolutionary Synthesis of NAND Logic: Dissecting a Digital Organism," in *Proceedings of the 2009 IEEE International Conference on Systems, Man, and Cybernetics* (Piscataway, NJ: IEEE Press, 2009), 3047–53; G. Montañez et al., "A Vivisection of the ev Computer Organism: Identifying Sources of Active Information," *BIO-Complexity,* no. 3 (2010): 1–6; W. Ewert, W. Dembski, and R. J. Marks II, "Climbing the Steiner Tree: Sources of Active Information in a Genetic Algorithm for Solving the Euclidean Steiner Tree Problem," *BIO-Complexity,* no. 1 (2012): 1–14; and W. Ewert, W. A. Dembski, and R. J. Marks II, "Active Information in Metabiology," *BIO-Complexity,* no. 4 (2013): 1–10.

11. N. Cheney et al., "Unshackling Evolution: Evolving Soft Robots with Multiple Materials and a Powerful Generative Encoding," GECCO '13, July 6–10, 2013, Amsterdam, The Netherlands.

12. www.biologic.org/stylus.

13. D. D. Axe, P. Lu, and S. Flatau, "A Stylus-Generated Artificial Genome with Analogy to Minimal Bacterial Genomes," *BIO-Complexity,* no. 3 (2011): 1–15.

14. D. D. Axe and A. K. Gauger, "Model and Laboratory Demonstrations That Evolutionary Optimization Works Well Only If Preceded by Invention: Selection Itself Is Not Inventive," *BIO-Complexity,* no. 2 (2015): 1–13.

Chapter 12: Last Throes

1. Charles R. Darwin, *The Origin of Species by Means of Natural Selection, or the Preservation of Favoured Races in the Struggle for Life,* 1st ed. (London: John Murray, 1859), 189.

2. "List of Scientific Bodies Explicitly Rejecting Intelligent Design," Wikipedia, last accessed May 7, 2016; https://en.wikipedia.org/wiki/List_of_scientific_bodies_explicitly_rejecting_Intelligent_Design.

3. "An Intelligently Designed Response," editorial, *Nature Methods* 4, no. 12 (December 2007): 983.

4. P. Shipman, "Being Stalked by Intelligent Design," *American Scientist* 93 (2005): 502.

5. E. Zuckerkandl, "Intelligent Design and Biological Complexity," *Gene* 358 (2006): 2–18.

6. "An Intelligently Designed Response," 983.

7. P. Ball, "What a Shoddy Piece of Work is Man," *Nature* online (May 3, 2010): doi:10.1038/news.2010.215.

8. Marshall Berman, "Intelligent Design: The New Creationism Threatens All of Science and Society," *APS News,* October 2005, 8.

9. G. Weissmann, "The Facts of Evolution: Fighting the Endarkenment," *FASEB Journal* 19 (2005): 1581–82.

10. Shipman, "Being Stalked by Intelligent Design," 502.

11. G. Petsko, "It Is Alive," *Genome Biology* 9 (2008): 106.

12. Hugo De Vries, *Species and Varieties: Their Origin by Mutation* (Chicago: Open Court Publishing, 1904), 4.

13. This sudden turn was discussed in chapter 1.

14. De Vries, *Species and Varieties,* 825–26. De Vries attributes the quote to a Mr. Arthur Harris, without citing a source. I have added emphasis to underscore the point.

15. W. Fontana and L. W. Buss, "'The Arrival of the Fittest': Toward a Theory of Biological Organization," *Bulletin of Mathematical Biology* 56, no. 1 (1994): 1–64. Emphasis added.

16. Fontana and Buss, "'The Arrival of the Fittest,'" 2.

17. A. Wagner, *Arrival of the Fittest: Solving Evolution's Greatest Puzzle* (New York: Current, 2014), 5 (emphasis in original) and 14.

18. Wagner, *Arrival of the Fittest,* 215–16.

19. Here, I should amplify a thought I alluded to in chapter 6, which is that the physical universe is itself a stunningly brilliant and beautiful invention. The point is that it's not an *inventor.*

20. E. V. Koonin, "The Cosmological Model of Eternal Inflation and the Transition from Chance to Biological Evolution in the History of Life," *Biology Direct* 2 (2007): 15.

21. In chapter 13, we'll see why the first assumption is incorrect. Once this is established, the existence or nonexistence of the multiverse becomes irrelevant.

Chapter 13: First World

1. Thomas Nagel, *Mind and Cosmos: Why the Materialist Neo-Darwinian Conception of Nature Is Almost Certainly False* (Oxford: Oxford Univ. Press, 2012), 31.

2. Nagel, *Mind and Cosmos,* 20.

3. Nagel, *Mind and Cosmos,* 25.

4. Nagel, *Mind and Cosmos,* 67.

5. Sarah Spaulding biography, Diatoms of the United States, http://westerndiatoms.colorado.edu/about/participant/spaulding_sarah.

Chapter 14: The New School

1. Roger Penrose does an excellent job of developing this point in *The Emperor's New Mind: Concerning Computers, Minds, and the Laws of Physics* (Oxford: Oxford Univ. Press, 1989).

2. "Grand Challenge: How Does the Human Brain Work and Produce Mental Activity?," in *From Molecules to Minds: Challenges for the 21st Century; Workshop Summary* (Washington, D.C.: National Academies Press, 2008), www.ncbi.nlm.nih.gov/books/NBK50989/.

3. J. M. Schwartz and S. Begley, *The Mind and the Brain: Neuroplasticity and the Power of Mental Force* (ReganBooks, 2003). Quoted from description on back cover.

4. I would be delighted to find that I've exaggerated the scarcity of biologists at research institutions who are free to work outside the confines of materialism. Like me, nearly everyone in this small collection of scientists must go outside the big establishment institutions to be free from their influence.

5. M.-J. Liu et al., "Biomechanical Characteristics of Hand Coordination in Grasping Activities of Daily Living," *PLoS One* 11, no. 1 (2016), doi:10.1371/journal.pone.0146193.

6. *PLoS One* Staff, "Retraction: Biomechanical Characteristics of Hand Coordination in Grasping Activities of Daily Living," *PLoS One* 11, no. 3 (2016), doi:10.1371/journal.pone.0151685.

7. *PLoS One* Staff, "Retraction."

8. E. Zuckerkandl, "Intelligent Design and Biological Complexity," *Gene* 358 (2006): 2–18.

9. William Bialek biography, Princeton University website, www.princeton.edu/~wbialek/wbialek.html.

10. Although scientists have for decades been able to make designed changes to the DNA carried by some organisms, this comes nowhere close to conceiving and designing a new form of life. In fact, having concluded that mind can't have a material basis, we conclude that it's impossible for humans to invent minds, and therefore equally impossible for humans to invent organisms *with* minds. We invent *thoughts* directly, some of these thoughts inspiring our material inventions. Minds, however, are neither material things nor thoughts; rather, minds are immaterial entities that *have* thoughts.

11. A common mistake is the thought that the new school can't be launched until it has physical descriptions of the origins of the various forms of life to replace the physical description offered by the old school. The insistence that there is nothing to be said about life apart from physical descriptions of physical processes is precisely the claim being challenged by the new school. By opening the door to the much more intellectually rich questions surrounding the *ideas* behind life, the new school launches itself in a completely different direction. The old question of what the various origins events would have looked like if they had been captured on video doesn't rank very high on the new school's priorities.

12. L. M. N. Meyer, G. C. Bomfim, and C. N. El-Hani, "How to Understand the Gene in the Twenty-First Century?" *Science and Education* 22, no. 2 (2011): 345–74.

13. M. B. Gerstein et al., "What Is a Gene, Post-ENCODE? History and Updated Definition," *Genome Research* 17 (2007): 669–81, http://genome.cshlp.org/content/17/6/669.full.pdf+html.

CREDITS

Grateful acknowledgment is given to the following for the use of their work in this book. All other illustrations are by Anca Sandu with art direction by Brian Gage.

Figure 3.3 (p. 32)	Laboratory of Molecular Biology, Cambridge, James King-Holmes/Science Photo Library (for illustrative purposes only).
Figure 3.4 (p. 33)	Six Nobel Prize Winners, NYPL/SCIENCE SOURCE/ Science Photo Library.
Figure 8.2 (p. 124)	Abraham Lincoln, public domain.
Figure 9.2 (p. 140)	Apollo 13 "mailbox," public domain.
Figure 10.1 (p. 166)	*Tavros 2*, Bgregson, used under the Creative Commons license CC BY-SA 3.0, via Wikimedia Commons.
Figure 10.4 (pp. 172–73)	Elements from "Localization of Membrane Proteins in the Cyanobacterium Synechococcus sp. PCC7942: Radial Asymmetry in the Photosynthetic Complexes," Debra M. Sherman, Tracy A. Troyan, and Louis A. Sherman, *Plant Physiology* © 1998 (used with permission) and from "Atomic-Level Models of the Bacterial Carboxysome Shell" Shiho Tanaka *et al.*, *Science* 319, 1083 (2008); DOI: 10.1126/science.1151458 (reprinted with permission from AAAS).
Figure 11.3 (p. 205)	NASA's evolutionary antenna, public domain.

Figure 11.4 (p. 207)	Soft robot image from Nick Cheney, Robert MacCurdy, Jeff Clune, Hod Lipson, "Unshackling Evolution: Evolving Soft Robots with Multiple Materials and a Powerful Generative Coding," *Proceedings of the 15th Annual Conference on Genetic and Evolutionary Computation,* © 2013 Association for Computing Machinery, Inc. Reprinted by permission.
Figure 11.6 (p. 211)	Chinese character and protein comparison, © 2008 Axe et al., *PLOS One,* CCY.
Figure 13.1 (p. 247)	Diatom shell, SEM, Steve Gschmeissner/Science Photo Library; diatom frustle, Steve Gschmeissner/Science Photo Library; diatom alga, SEM, Steve Gschmeissner/Science Photo Library; diatom, SEM, Steve Gschmeissner/Science Photo Library.
Figure 13.2 (p. 248)	*Ellerbeckia arenaria* close-up, L. Bahls (2014) in *Diatoms of the United States,* retrieved February 23, 2016, from http://westerndiatoms.colorado.edu/taxa/species/ellerbeckia_arenaria.
Figure 14.3 (p. 264)	"Atheist Bus Campaign Launch," Zoe Margolis, used under the Creative Commons license CC BY 2.0, via Wikimedia Commons.
Plate 3 (insert p. 2)	Merlet's scorpionfish, Georgette Douwma/Science Photo Library; mandarinfish, Dollar Photo Club © Olga Khoroshunova; Siamese fighting fish, © visarute angkatavanich / 500px; fringehead fish, Chris Newbert/Minden Pictures/National Geographic Creative; red-lipped batfish, NatalieJean/Shutterstock; giant stargazer, Joe Belanger/Shutterstock; fangtooth, Dante Fenolio/Science Photo Library.

INDEX

Index

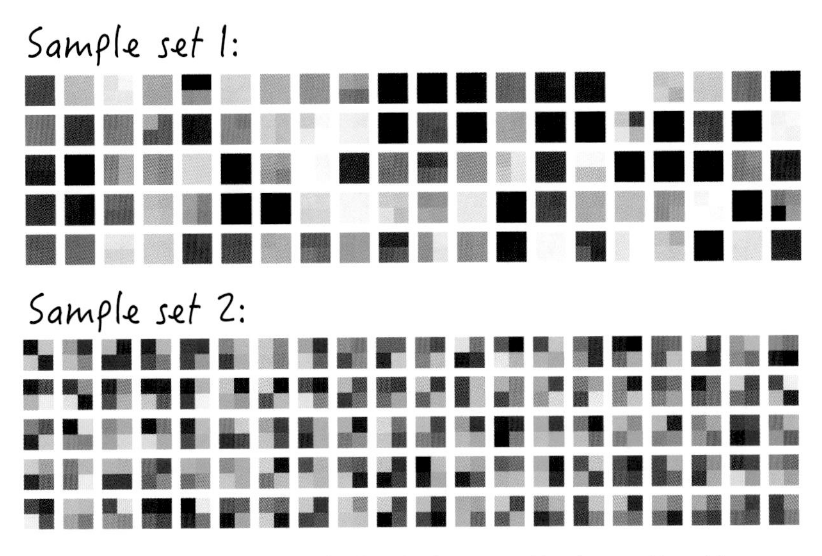

Plate 1 The visual effects of bottom-level pixel coherence and incoherence. To avoid compression artifacts, uncompressed TIFF files were used for all photos.

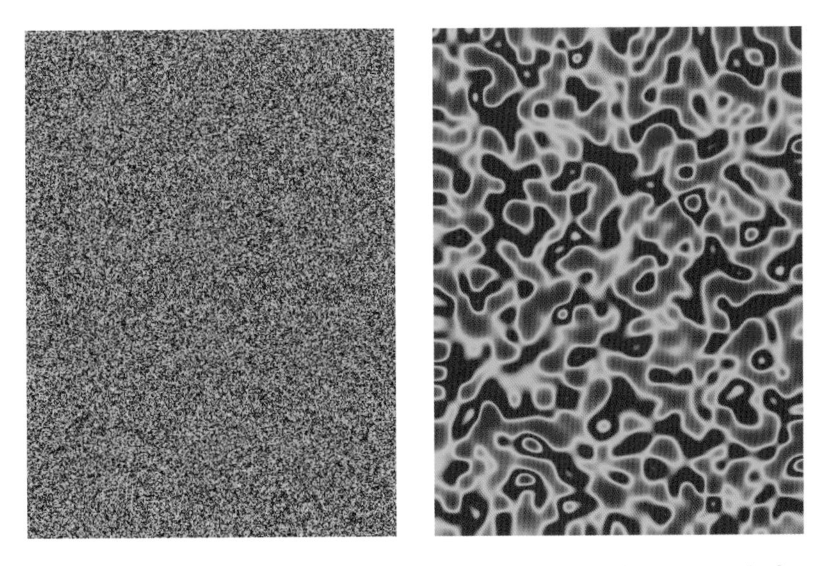

Plate 2 A random image (*left*) and a nonrandom image (*right*), obtained by processing the first image with the *Mathematica* commands *ImageResize* and *Colorize*.

Plate 3 Fish by genre. *Top row:* Merlet's scorpionfish (*Rhinopias aphanes*) and mandarinfish (*Synchiropus splendidus*). *Second row:* Siamese fighting fish (*Betta splendens*). *Third row:* fringehead fish (*Neoclinus blanchardi*) and red-lipped batfish (*Ogcocephalus darwini*). *Bottom row:* giant stargazer (*Kathetostoma giganteum*) and fangtooth (*Anoplogaster cornuta*). Unlike the other fish shown, the Siamese fighting fish have been bred in captivity in order to bring out their full potential. Because of this, some may say they're less compelling as a demonstration of the emotional connection between us, their observers, and God, their creator. To me this only makes them more compelling, in that it seems we've been invited to participate, in a very small way, in the creative process itself.